Communications Handbook

Communications Law Handbook

General Editor

Mike Conradi
Kemp Little LLP

Bloomsbury Professional

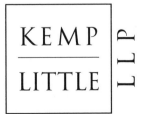

Bloomsbury Professional
Maxwelton House
41–43 Boltro Road
Haywards Heath
West Sussex
RH16 1BJ

© Chapters 1–4, 6 and 8 Bloomsbury Professional Ltd 2009

© Chapter 5 George Ritchie 2009

© Chapter 7 Allen & Overy LLP 2009

British Library Cataloguing-in-Publication Data
A CIP Catalogue record for this book is available from the British Library.

ISBN 978 1 84766 311 5

Typeset by Phoenix Photosetting, Chatham, Kent
Printed and bound in Great Britain by CPI Antony Rowe, Chippenham, Wiltshire

Contents

Contents

Table of Statutes

Table of Statutory Instruments

Table of European Legislation

Table of Cases

Table of Cases

Chapter 1

Introduction

Mike Conradi
Kemp Little LLP

Introduction

The regulation of communications in the UK is a fascinating subject largely because it involves the complex interplay of so many factors. As the long heralded era of 'convergence' finally seems to be arriving, each year brings an array of new types of services and products each of which raises its own issues in terms of regulation. Even leaving aside the technological aspect, communications regulation is uniquely interesting as it involves issues of competition law, data protection, content rules, wireless telegraphy and public law. All of these various areas of law have particular application to the telecoms sector and the aim of this book is to explain how these all fit together. The 'Practical Issue' boxes throughout highlight, at the time of writing, some of the most interesting current topics but new ones arise all the time.

The 2003 Regulatory Framework Regime

Fundamental to this book is an understanding of the five directives collectively known as the 2003 regime. These were the result of a comprehensive review of the telecoms regulation which the European Commission undertook in the years leading up to 2002. To understand just how significant these directives were, it is first necessary to understand something about how regulation worked before 2003 in the UK.

Before 2003 it was an offence under the Telecommunication Act 1984 for a person to 'run a telecommunications system' unless authorised to do so by means of a licence granted under the terms of that Act. There was a very complex system of individually issued licenses whereby each licence contained a different set of restrictions and different licences were needed in order to provide different types of services. Over time OFTEL (then the sector regulator) had made various changes to its standard licensing templates and the result was that different operators had slightly different terms applicable to them. BT had an entirely different licence from the others. Moreover a system of 'Class Licenses' permitted entire categories (or 'classes') of users, including consumers in their own homes, to run a telecom system subject to the constraints set out in those licenses.

All-in-all it was a complex and confusing system without any overriding logic to it. Whilst a fantastic arrangement for telecoms lawyers it was clear that the system was generating obstacles to competition in the market.

Introduction

The 2003 regime changed all this, not just in the UK but throughout the EU, by removing all of the old system and replacing it with a new framework. This consisted of the following directives:

- Directive 2002/21/EC—the 'Framework Directive', which creates a framework within which the other directives sit.

- Directive 2002/20/EC—the 'Authorisation Directive', which harmonised and simplified the authorisation regime.

- Directive 2002/19/EC—the 'Access Directive', which dealt with access to an interconnection of networks and associated facilities.

- Directive 2002/22/EC on Universal Service—which established various consumer protection and end user rights and aimed to ensure the availability of a basic set of services throughout the European Union.

- Directive 2002/58/EC—the 'E-Privacy Directive', which concerns confidentiality and privacy in information generated by communications services.

Each of these is described in more detail below.

The Framework Directive 2002/21/EC

This directive contains (at Article 2) the key definitions of:

- 'Electronic Communications Network' (ECN)—a transmission system which permits the conveyance of signals irrespective of the type of information conveyed.

- 'Electronic Communications Service' (ECS)—a service normally provided for remuneration which consists wholly or mainly in the conveyance of signals on an electronic communications network including transmission services in networks used for broadcasting but excluding services providing, or exercising editorial control over, content.

The concept of ECNs and ECSs is fundamental to the rest of the 2003 regime.

The Framework Directive goes on to require each member state to set up an independent National Regulatory Authority (NRA) and to ensure that there are effective mechanisms in place to appeal any decisions made by the NRA. NRAs are given a duty to promote competition, and to remove obstacles to service provision. They also must ensure there is no discrimination in the treatment of undertakings providing ECNs or ECSs.

Importantly the Framework Directive also establishes what is known as the 'market definition' and the 'market analysis' procedures. Article 15 sets out the market definition procedure. It says that after public consultation the Commission should adopt a 'recommendation on relevant product and service markets' identifying those product and service markets within the electronic communications sector, the characteristics of which may be such as to justify the *ex ante* imposition of 'regulatory obligations' as set out in the other Directives. The history of the two recommendations (to date) issued by the Commission and their contents is discussed at Chapter 4 of this book. NRAs then have an obligation to 'take the utmost account' of the recommendation in defining in their own jurisdictions a set of markets which may be suitable for *ex ante* regulation. The next stage is the 'market analysis' procedure as set out at Article 16. This requires NRAs to carry out a detailed analysis of the markets which they have identified in accordance with the market definition procedure. Where an NRA concludes that a market is effectively competitive it is obliged not to impose any specific *ex ante* regulatory obligations (relying instead on the normal *ex post* principles of competition law to ensure a competitive outcome). By contrast, where they determine that a relevant market is *not* effectively competitive then the NRA has an obligation to identify one (or more) undertakings with 'significant market power' (SMP) on that market and then it must impose remedies on the SMP operator.

'Significant market power' is defined as a 'position equivalent to dominance'. The difference then between the general competition law concept of 'dominance' and the telecoms-specific definition of SMP is that, in concept, no consequences follow simply from the fact that a business is dominant in a market. Instead, the legal consequences follow only if a dominant player abuses its dominant position (such as by requiring customers purchasing products in the non-competitive market to also purchase separate products from the competitive part of the market). By contrast the simple fact that an operator has SMP in one of the relevant markets *must necessarily* lead to the imposition of remedies on it by the NRA whether or not the operator concerned has in fact done anything to reduce or lessen competition in the market or otherwise to abuse its position.

The Directive also contains an obligation on NRAs to withdraw any *ex ante* regulation from markets which are deemed to be effectively competitive. In this way the Framework Directive aims to ensure that there is a logical and consistent process followed in order to determine which regulations apply to whom in the various market sectors.

Access Directive (2002/19/EC)

This directive sets out the basis on which operators access one another's networks. Importantly, it contains an obligation on the operators of any public network to negotiate interconnection with one another for the provision of publicly available ECSs. This obligation, contained in Article 4, is arguably one of the most significant provisions of the entire 2003 regime. This is because it imposes an obligation even on businesses that do not have significant market power and it is via this mechanism that disputes between any two operators can be referred to the relevant NRA.

Article 5 gives NRAs the power to impose an obligation 'to the extent that it is necessary to ensure end-to-end connectivity'. Again this is irrespective of significant market power and in the UK OFCOM has exercised this power by imposing a condition requiring BT to purchase various wholesale services from any public ECN that reasonably requests it in writing. This condition was imposed in 2006 after OFCOM decided that the previous 'guidance' on end-to-end connectivity was not sufficient. The issue has become particularly important in the UK in the context of mobile call termination. Over a series of legal proceedings the mobile operator H3G ultimately failed in its argument that it should not be deemed to have significant market power in the market for terminating calls on its own network, due, it said, to BT's 'countervailing buying power'. Their argument ran that BT was such a large purchaser of termination services that its buying power was so strong as to mean that H3G did not in fact have a 'position equivalent to dominance' even in respect of the market for calls to its own subscribers. After a lengthy set of legal challenges the ultimate conclusion reached was that the end-to-end connectivity obligation imposed by OFCOM negated this countervailing buying power because BT could not make a credible threat to cut H3G's customers off. More details on this are contained in Chapter 4.

The other significant part of the Access Directive describes the market review process in more detail and says, at Article 8, that where an operator is designated as having significant market power in a specific market then the national regulatory authorities must impose at least one of a set of 'remedies' (or 'SMP conditions') as set out in the directive. These are obligations of transparency (Article 9), non-discrimination (Article 10), accounting separation (Article 11), access obligations (Article 12) and price control and cost accounting obligations (Article 13). Article 8 also says that NRAs may not impose these obligations on operators that have not been designated as having SMP.

Authorisation Directive (2002/20/EC)

This Directive abolished the old licensing system and replaced it with a system allowing anyone to provide an ECN or ECS without the need to obtain a specific licence, subject only to a set of conditions attached to a 'general authorisation' which must apply equally to all operators in the market place.

Under Article 3 NRAs in each Member State do have the power to impose a notification requirement under which it is necessary to tell the NRA before providing an ECN or ECS, but this a notification only—the NRA has no power to approve or refuse authorisation. In the UK OFCOM has, at present decided not to impose such a notification requirement. As a consequence any business may provide electronic communications networks or service without any need to notify the regulator at all provided only that they comply with the general authorisations (described in detail at Chapter 3).

The Authorisation Directive does permit individual licences in the case of radio spectrum but only where this is 'necessary' (bearing in mind the risk of interference) and any such licences must be granted in accordance with an open transparent and non-discriminatory procedure. The radio spectrum authorisation regime in the UK is discussed in more detail in Chapter 7.

The Directive contains an annex which sets out an exclusive list of the conditions which can be attached to a Member State's general authorisation. These are legally separate from any specific obligations which can be imposed on individual providers either because they have significant market power (following a market review in accordance with the Access Directive) or because they have had conditions imposed on them by other provisions of the 2003 framework (such as the end-to-end connectivity obligations discussed above).

Universal Service Directive (2002/22/EC)

This Directive sets out a set of services which should be available to all end users irrespective of location at an affordable price. The most important of these is set out at Article 4—'provision of access at a fixed location'. This says that all reasonable requests for connection at a fixed location should be met and that the connection should be capable of allowing local, national and international calls, fax communications and 'data communications at data rates that are sufficient to permit functional internet access, taking into account prevailing technologies used by the majority of subscribers and technological feasibility'. This last part means that some form of internet access should be available across the European Union although there is of course some considerable debate as to what this means. In the UK OFCOM has implemented Article 4 by obliging BT to offer this set of services at a standard price to anyone in the country where installation of a new line costs £3400 or less. Where installation will cost more than this BT may require the customer to pay the excess cost plus its standard connection charge[1].

At the time of writing OFCOM's guidance suggests that the minimum speed required for 'functional internet access' is at least 28.8 Kbytes per second (i.e. a very slow speed, equivalent to the standard dial-up internet access speed attainable in the mid-1990s). However there have recently been proposals to expand this obligation so that the minimum speed available would increase roughly one hundred times to 2 mb/s[2].

Other services which should be available to all end users include directory enquiries, pay phones (if needed), and services to enable disabled users to access communications.

The Directive enables Member States to introduce a mechanism to compensate operators on whom universal services obligations are placed. Typically this would take the form of a universal services fund into which any ECN (or ECS) provider would contribute. However in the UK the current view is that no such fund is necessary—BT obtains a benefit from being known as the universal services provider which exceeds the cost of this provision. If however the cost of providing universal services were to increase significantly (as it would if the minimum broadband speed were to be increased significantly) then a change may be made.

The Directive also contains a number of consumer protection measures which apply to any provider of access to a public telephone network. These include an obligation to include certain specific provisions in contracts, the publication of quality of service information and activating the EU-wide emergency services number (112).

The Privacy and Electronic Communications Directive (2002/58/EC)

This Directive applies to the processing of personal data in connection with the provision of a public ECN and is discussed in Chapter 2. It contains an obligation on service providers to notify their customers of security risks (but not, at present, of particular security breaches). It contains further provisions intended to ensure the confidentiality of communications and a provision requiring service providers to

---

1   In fact this obligation applies not only to BT but also to Kingston Communications. Kingston provide telecom services in Hull an area where, traditionally, BT had no network at all. Throughout this book many obligations that are imposed on BT are also imposed on Kingston. The universal service obligation can be found at www.ofcom.org.uk/consult/condocs/uso.
2   This proposal was contained in the government *Digital Britain – Interim Report* published in 2009. No details were given as to how this additional obligation would be funded. See www.culture.gov.uk/what_we_do/broadcasting/5631.aspx.

anonymise or delete traffic and location information when it is not needed. It also contains rules on unsolicited emails.

Contents of the book

The remaining chapters of this book explore in some detail how these Directives have been implemented in the UK and discuss, where relevant, the forthcoming reform process and the changes this is likely to bring.

Although the book concentrates only on the UK it may nevertheless be helpful when considering regulation in other countries. This is because it is still the case the UK, and OFCOM in particular, frequently set a lead in telecoms regulation which is subsequently followed by others. The principle example of this, of course, was the creation of Openreach discussed in further detail in Chapter 5 but other examples include the UK being the first country to conduct a market review process and to remove most price controls from its incumbent operator (discussed in Chapter 4).

The following paragraphs describe at high level the contents of the remainder of the book:

Chapter 2—Data Protection and Privacy

This chapter covers all aspects of privacy in communications. It discusses the rules on data protection, interception and data retention with reference to the Data Protection Act 1998, the Regulation of Investigatory Powers Act 2000, the Privacy and Electronic Communications Regulations 2003 and to other legislation. Rather than present a complete description, though, it focuses on aspects relevant to communications. The chapter includes a helpful diagram showing one potential structure for an 'intra-group agreement' (IGA) which is designed for use in the context of transfers of personal data between a large number of companies within a single group. An 'issues box' outlines the law on whether companies can record incoming calls (a topic the editor is frequently asked to advise on)

Chapter 3—OFCOM and the General Conditions of Authorisation

This chapter describes each of the regulatory requirements (or 'general conditions') which apply to any provider of electronic communications networks or services in the UK. It explains the rationale behind the evolution of some of them and provides a guide to any new entrant or competing service provider active in the market today or contemplating launching a new service.

Chapter 4—Specific Conditions

In contrast with Chapter 3, which describes conditions applicable to all players in the marketplace, Chapter 4 describes the various types of conditions which can be imposed on specific businesses (usually called 'undertakings' in the directives). The main ones, of course are the conditions imposed as a consequence of a finding of significant market power in accordance with the market analysis procedure set out in the Access Directive but there are others as well. Specifically universal service conditions, access related conditions and, in theory, 'privileged supplier' conditions. The chapter describes how each of these work and includes some very helpful tables showing the current status of the various market reviews, OFCOM's conclusions in respect of them and the SMP conditions which have been imposed.

Chapter 5—Strategic Review and Creation of Openreach

This chapter describes OFCOM's 2005 strategic review of telecoms and the process which led to functional separation and the creation of Openreach. It explains the structure of the undertakings given by BT to OFCOM (in lieu of a reference to the Competition Commission) under which Openreach was set up and describes the relationship between Openreach and the rest of BT. The chapter provides an invaluable guide to anyone faced with the daunting prospect of working through the undertakings to answer any specific question—content which is not available anywhere else. It includes a description of the changes proposed as part of OFCOM's 'Future Broadband' consultation under which Openreach will go ahead and build a 'next generation access network' (i.e. upgraded links between a customer's home or office and the local exchange enabling much faster broadband speeds).

Chapter 6—Content Regulation

This chapter covers the regulation of content with a particular focus on new media platforms such as the internet and mobile phones. It describes briefly the history of content regulation in the UK and talks about a number of codes of practice which are in place on a voluntary basis and published by various industry groups in the UK. It includes a description of PhonepayPlus, the premium rate line content regulator, and its powers to regulate premium rate telephone numbers. It also includes further detail on the forthcoming Audio Visual Medial Services (AVMS) directive due to be implemented in the UK by the end of 2009.

Chapter 7—Wireless Telegraphy

This chapter describes the regulation of radio frequency spectrum and its allocation in the UK. It covers the Wireless Telegraphy Act 2006 and sets out the processes by which spectrum is now allocated. It also describes the European radio spectrum policy and explains how this relates to the UK. It includes an explanation of when a specific wireless telegraphy licence is needed or when, by contrast, reliance can be placed on some other legal basis for the use of spectrum.

Chapter 8—Powers and Duties

This chapter concentrates on the legal powers of OFCOM and its duties mostly under the Communications Act 2003 although also under the Competition Act 1998. It looks at some of the recent cases decided by the Competition Appeals Tribunal and explains how these give guidance on the way in which OFCOM should make decisions or resolve disputes. It includes a table setting out every Competition Act complaint which has been decided by OFCOM since its inception, with brief information and details of the outcome.

The chapter also covers some of the relevant criminal offences as well as the powers held in this context by the UK government—currently exercised by the Department for Business Enterprise and Regulatory Reform (BERR).

Forthcoming Reforms

At the time of writing the EU is close to finalising a package of reforms to the 2003 regime (the 'Framework Review' package)[3]. The European Commission began this review in the summer of 2006 with detailed proposals first published in November 2007. The Framework Review consists of the following:

Change to the recommendation on product and service markets

In November 2007 the Commission published its second recommendation on the relevant product and service markets susceptible to *ex ante* regulation (as required by the Framework Directive). This reduced the number of markets which NRAs were expected to review down from 18 to just seven. At the retail level the only market remaining is 'access at a fixed location'. The remaining six markets are all wholesale and mostly relate to fixed (rather than mobile) markets. The new recommendation was effective immediately on publication in November 2007 and we can therefore expect NRAs to remove any *ex ante* regulation in markets which no longer form part of the Recommendation over the next few years. For example the wholesale market for access to mobile networks—which is the market in which MVNOs (mobile virtual network operators) buy access—was included in the original recommendation but has now been removed. This means that those countries which imposed access obligations on their mobile operators are likely to lift those obligations shortly.

The new Recommendation is discussed further at Chapter 3 but it is worth noting here that the continued distinction between fixed and mobile markets may not be sustainable—it is clear throughout the EU that mobile calls are increasingly used in substitution for fixed ones and therefore we can foresee the day, perhaps in a third recommendation, where many of these distinctions may be removed.

The 'Citizens' Rights Directive' (a draft directive amending the Universal Services Directive, the E-privacy Directive and Regulation 2006/2004 on Consumer Protection Cooperation)

This is the first of two directives which themselves amend directives forming part of the 2003 regime. The changes they propose are discussed below but the stated aim is to strengthen certain consumer and user rights, to ensure that electronic

---

3   On Wednesday 6th May 2009, the European Parliament (EP) voted on the Second Reading of the Framework Review package. This was expected to be a final rubber-stamp but the EP voted to re-insert an amendment stating that a judicial ruling was required before an end-user's internet access could be blocked. This was aimed at preventing some national governments' proposals for dealing with copyright and piracy issues by compelling ISPs to withdraw services from consumers in certain circumstances. As a result the future of the whole of the Framework Review package is, at time of writing, uncertain though it is anticipated that ultimately the changes described in this section will become law.

communications are trustworthy, secure and reliable and to provide a high level of protection for individuals' privacy and personal data.

The 'Better Regulation' Directive (a draft directive amending the Framework Directive, the Access Directive and the Authorisation Directive)

As above this directive proposes amendments to directives forming part of the 2003 regime. The stated aim is to improve the effectiveness of the Directives being amended, to reduce the administrative resources needed for implementing economic regulation and to make access to radio frequencies simpler and more efficient.

A draft regulation establishing the Body of Regulators for Electronic Communications

The aim of this regulation is to set up a new pan-European regulatory body. This would establish an independent expert body that would make an effective contribution to furthering the completion of the internal market by assisting the Commission and the national regulatory authorities in implementing the EU regulatory framework.

A new GSM Directive

This is discussed below, but in brief this new directive will allow GSM spectrum to be freed up for other uses.

Key proposals

The main changes which are expected when these proposals become law are as follows:

- **Creating a new pan-European regulatory body.** The European Commission's original proposals were for the creation of a new 'market authority' with a role, amongst other things, in authorising number ranges and spectrum for pan-EU services at an EU level. The intention was that the market authority would replace the current European Regulators Group (ERG) and also takeover the work of the European Network and Information Security Agency (ENISA). After discussion within the EU this went through several modifications and name changes. At time of writing, though, it seems likely that it will be called the Body of European Regulators for Electronic Communications ('BEREC'). It will have a lesser role than was originally envisaged for the market authority—less power, and, importantly, will be funded by the Member States and not by the Commission. It's final role and powers are yet to be finalised but it seems clear that it will focus on telecoms regulation and have no role in relation to spectrum or network security, meaning that ENISA will not be merged with it (at least until 2013, when this will be reviewed again). Specifically BEREC will:

- be comprised of the heads of each of the NRAs and a representative from the European Commission;
- replace the ERG, offering reflection and debate as well as to advice to the Commission, the European Parliament and the European Council;
- not itself have legal personality, though there will be an autonomous 'Office' with legal personality that will provide professional and administrative support. The Office will be funded by the Community budget and by voluntary national contributions;
- have a clearly defined role in drawing up regulatory best practices to be disseminated among NRAs;
- issue opinions, recommendations and guidelines of which the NRAs and the Commission will be required to take the 'utmost account';
- conduct its activities with a high level of transparency. It will be required to publish both its annual programme and report to the Parliament and Council annually on its activities.

- **Spectrum.** Introducing greater spectrum liberalisation and harmonisation based on the principles of technology and service neutrality, market-led authorisations and secondary trading. These proposals may well turn out to be the most significant of all the reforms to be made. Harmonisation of spectrum seems likely to facilitate the creation of more cross-border pan-European telecoms services, and the principles of technology and services neutrality are intended to ensure that the market, rather than regulatory pressure determines the most appropriate use of spectrum. Similarly, allowing more secondary trading of spectrum (where it is licensed rather than made available under a general authorisation) seems likely to lead to significant improvements in the efficiency of spectrum use. In its original proposal, the Commission proposed changes that would secure a more important role for itself in planning and coordination of spectrum policy however it now seems likely that Member States will retain ultimate control over national spectrum, albeit under harmonised spectrum trading rules. The reforms in this area will bring a much stronger emphasis on flexibility in spectrum use, allowing operators to introduce innovative technologies and services quickly. The commission claims that this has the potential to generate an estimated additional 0.1 per cent of GDP per annum, and that the new flexibility will in particular allow maximised use of the 'digital dividend', the radio spectrum freed as a result of the switchover from analogue to digital TV[4].

- **Boosting 3G mobile services:** The Framework Review is intended to pave the way for the swift adoption of the new GSM Directive. The new Directive will allow any service, starting with 3G and extending later to other new technologies, to operate in the GSM band which is currently reserved exclusively for GSM services. This will, the Commission says, lead to industry savings estimated at €1.6 billion in capital costs for a **single** Europe-wide network, and enable faster roll-out of full 3G coverage[5]. This will boost the take up of 3G in the EU, from the existing 3G customer base which was estimated at over 90 million in December 2008. The penetration rates of 3G are currently the highest in Italy, Austria, Sweden and the United Kingdom where they exceed 20 per cent of the total subscribers.

---

4 See press release at http://europa.eu/rapid/pressReleasesAction.do?reference=MEMO/09/219&form at=HTML&aged=0&language=EN&guiLanguage=en.
5 *Ibid.*

- **Remedies.** Under current rules the Commission may veto the decision of an NRA on market definition if the NRA defines a market in a way which differs from those set out in the current recommendation' on that subject. It can also veto an NRA's finding that a particular operator has SMP in a market. The European Commission cannot, however, veto an NRA's choice of remedies to address the presence of SMP. The proposal, then, was to give the Commission a new power to veto any remedies proposed by an NRA, following a market review, where they have made a finding that a market is not competitive (and therefore that at least one player has significant market power). This would have been a new power aiming to address one of the problems which the commission identified during the Framework Review—an inconsistency in remedies proposed across Member States. However in what seems likely to be the final form of this proposal the Commission will not have a veto power over remedies proposed by NRAs. Instead NRAs will be obliged to take 'utmost account' of the Commission's, or BEREC's, opinion and if both the Commission and BEREC have 'serious doubts' then the NRAs must co-operate with them both to identify the 'most effective measure'. In effect the result is that the Commission has the power to delay the imposition of a remedy with which it disagrees, and to make the NRA justify its decision in writing, but no veto.

- **Functional separation.** A further change to be introduced will set up a separate process whereby, in 'exceptional circumstances' NRAs will have the power, subject to prior approval from the European Commission, to impose a remedy of functional separation on operators with significant market power. This measure was strongly opposed by some incumbent operators but nevertheless seems likely to form part of the final amendments. Chapter 5 of this book discusses in detail how functional separation has been achieved in the UK. It is worth noting, though, the peculiar situation in the UK—where OFCOM had available to it, in addition to the powers granted under legislation implementing the 2003 framework, the threat of a review under the Enterprise Act to the Competition Commission. It may be that this situation, being unique to the UK, means that it would be more difficult to impose functional separation on an unwilling incumbent in any other country.

- **Strengthening the NRA.** The proposals in this area include a requirement on Member States to ensure that the head of the NRA can only be removed from office if he or she no longer fulfils the conditions required for the performance of their duties or if they have been guilty of serious misconduct. Other proposals also aim to strengthen the independence and power of the NRA. For example the proposals include a clarification that interim measures may only be granted to suspend an NRAs decision on appeal if there is an urgent need to suspend the effect of that decision to prevent serious and irreparable damage.

- **Privacy and Data Protection.** Providers of publically available electronic communications services seem likely to be required, under a new provision, to notify their NRA and, in some circumstances any affected users of any breaches in personal data security. This is a potentially significant change which would, under the current drafting, apply to any telecoms operator or internet service provider and it has attracted some considerable attention. It is worth noting, though, that this proposal is significantly less widely applicable than similar laws in certain states in the USA (eg California) which require public notification of security breaches for a much wider range of businesses. A proposal to extend the notification requirement to any providers of 'information society services' (including online banks, online retailers and online pharmacies) looks likely to be rejected. Member States should give data protection authorities power to set detailed rules and issue detailed instructions concerning the circumstances where notification is required. Other proposed changes in this area would impose more general obligations on network providers to ensure that personal data can be accessed only by authorised personnel for legally authorised purposes, to protect personal data against accidental or unlawful destruction, alteration or processing and to implement a security policy with respect to the processing of personal data.

- **Consumer Protection.** The proposals also include various measures to improve the transparency of information provided by service providers to consumers, particularly on the terms of supply and on their tariffs. There is also a proposal to allow any subscriber to 'port' their fixed or mobile telephone number within a period no longer than one working day. Other measures include improvements to the implementation of the 112 emergency number, improvements to access to free phone numbers called from abroad and empowering NRAs to impose minimum service quality requirements based on standards drawn up at an EC level. Finally, in this category, the proposal suggests an amendment to the E-privacy Directive which would give legal persons with a legitimate interest in combating the sending of unsolicited commercial emails (know as 'spam') the right to take legal action against spammers in civil proceedings. It also suggests that unsolicited emails disguising or concealing the identity of the sender should be banned.

- **Regulatory Holidays, Next Generation Networks and Infrastructure Sharing.** There has been some debate in this area on the principle of 'regulatory holidays'. This concept refers to a decision by an NRA to incentivise significant investments in networks by allowing (usually incumbent) operators a period of time after completion of their investment before they will be required by regulation to give access to the new network infrastructure their competitors. The European Commission and Parliament has confirmed its strong opposition to any such concept despite intensive lobbying from some countries (especially Germany), and has stated that NRAs have the power to oblige dominant players to share facilities such as buildings, entries to buildings, building wiring, masts, antennae, towers and ducts. Instead NRAs should incentivise investment by ensuring that any access obligation imposed on incumbent operators takes appropriate account of the risks incurred in building the infrastructure in the first place. It seems clear, then, that the concept of regulatory holidays will be opposed—the argument against being that if the price for access is set appropriately then the incentives for investments should remain. Indeed, in the UK, BT is currently building its core Next Generation Network (NGN, also known as the 21st century network or 21CN) at a total cost of some £10bn. It is doing this despite the absence of any 'regulatory holiday' provision and has stated that it expects to save £1bn annually through this investment.

What the book does not cover

Finally, a few words setting out what this book does *not* cover:

- The book is intended to provide a detailed discussion only of UK law and regulation. Although there are a few references to other countries, these are usually simply by way of illustrating a point about the UK regime. The book does not cover communications law in other countries.

■ Generic data protection rules are not covered by this book, though Chapter 2 does of course cover the specific data protection and privacy rules as they relate to communications services specifically. There are other publications which cover data protection more generally for those looking for this.

Although chapter 6 covers content regulation (and specifically those which will shortly apply to new media services) and chapter 4 includes a brief description of the access condition which has been imposed on Sky in respect of their platform, the book does not include a detailed description of the broadcasting regulations and the rules around operating a conventional broadcast television channel (via a 'television licensable content service' licence). This is because a detailed description of the regulation and law in this area would be a significant addition to the book, and is unlikely to be of interest to readers concerned with telecommunications.

Chapter 2

Privacy in Communications— Data Protection, Interception & Data Retention

Cameron Craig
DLA Piper LLP

1 Data Protection Act 1998

1.1 Introduction

In an era of rapid technological advances, intangible assets such as personal data have become a valuable commodity. Businesses have been quick to realise that personal information about an individual can be exploited in a number of ways to further their position in the marketplace. Although many IT and telecommunications businesses are reaping the rewards of the 'information economy', it is vital that the industry sector appreciates the risks involved in the use of personal data.

It has long been recognised that individuals have a right to privacy of their personal information. To that end, governments have sought to regulate and supervise the processing of personal data. The main forum for development of regulatory protection has taken place in Europe. However, in recent years, the topic has gained international appeal and data protection has thus become a global issue. The UK data protection regime has been borne out of European initiative and is structured upon the 1995 EU Directive (95/46/EC). The Directive was implemented in the UK through the Data Protection Act 1998 ('DPA'). The objective of the DPA is to give individuals the right to know what information an organisation holds about them and to provide a framework to ensure personal information is handled properly. The Act is often seen as the lynchpin of the UK data protection regime, nevertheless, it is important to appreciate that its function is performed within a myriad of other supporting legislative instruments. This chapter will explore the data protection regime as a whole in both a domestic and international context.

1.2 History of the DPA

Proposals for a regulatory framework began in 1979 with the European Parliament's Legal Affairs Committee.

The Committee was charged with investigating how best to safeguard the rights of individuals against the dangers posed by automatic data processing. In its report[1], the Committee identified both a technical and legal threat to adequate protection of personal data. On a technical level, it recognised that there was an increasing use of computing methods and computer databases by which organisations were collecting and storing data. The legal issue was that national provisions in the Community states were diverse in their approach to data protection laws. Accordingly, the Committee proposed a Community Direction on the harmonisation of legislation on data protection to provide citizens of the Community with maximum protection. In the meantime some member states began to implement their own data protection laws. In a bid to prevent disparity between the states, the European Commission drew up the first proposal on the directive. Directive 95/46/EC ('Directive') was adopted on 24 October 1995 with a requirement that it would be implemented in each member state by October 1998.

In the UK, the Directive was transposed into the Data Protection Act 1998 and received the Royal Assent in July of that year. Although one of the objectives of the Directive was to achieve harmonisation throughout the EU, its actual provisions allowed a fair degree of flexibility in the implementation process. As a result there are differences[2] in the way the Directive has been implemented in the member states and significant differences in the way in which the law is enforced by national data protection regulators.

A large number of guidance notes and codes of conduct are available through the Information Commissioner's Office (ICO) website (www.ico.gov.uk) and the Ministry of Justice which are helpful in interpreting and understanding the DPA. However, there can be no guarantee that compliance with ICO or Government guidance or codes will always ensure lawful processing of data and ultimately it will be the courts' interpretation that will be used as the benchmark for compliance.

1.3 DPA definitions

The DPA is a complex document which comprises six parts and a number of schedules. Before considering the scope of the Act, it will be helpful to set out the definitions of the technical terms that are not only used throughout the Act but are also used in common parlance of data protection;

- '*data controller*' means a person who (either alone or jointly or in common with others) determines the purposes for which and the manner in which any personal data is, or is to be, processed;

- '*data processor*' means a person who processes the data on behalf of a data controller. In simplified terms, where an organisation or person begins to determine the purposes for and manner in which data is processed it then becomes a 'data controller';

---

1   Report on the protection of the individual in the face of technical developments in data processing (1979–80 Eur Parl Doc (No 100) 13 (1979).
2   For example some member states, including Italy, have implemented law which applies to corporate data as well as data relating to individuals.

- *'personal data'* is data relating to living individuals who can be identified from the data, or from the data and other information which is in the possession of, or is likely to come into the possession of, the data controller;

- *'processing'* of data includes obtaining, recording or holding the data and disclosure of the data as well as a large variety of other acts in relation to data.

There has been feverous debate in recent years over the exact remit of the definitions used in the DPA. In particular, the definition of 'personal data' has caused great concern between practitioners and academics alike. The issue will be discussed later on in this chapter.

1.4 Obligations under the DPA

The DPA imposes obligations on the data controller who undertakes the processing of personal data.

A fundamental requirement is that data controllers process personal data in accordance with the eight principles[3] set out at Schedule 1 of the DPA. Processing is defined very broadly and includes the obtaining, recording, holding, using, disclosing and even erasing of data.

A further key additional obligation imposed on data controllers is the duty to register with the ICO. A data protection registration, or notification as it is now known under the DPA, is required for each entity which acts as a data controller, so a 'group notification' is not possible. In accordance with section 18 of the DPA, the data controller must provide the regulator with information regarding his identity; the type of data to be processed; the purpose for processing the data; the recipients of any data and security measures taken for the protection of data. Notification is an inexpensive procedure in the UK and can be done with a relative amount of ease depending on the size of the organisation. Currently, a hard copy form has to be sent to the ICO with the standard fee of £35, however, this may be subject to change in the future. With effect from 1 October 2009 a two tier fee structure will be in operation. Public authorities with 250 or more employees and private organisations which have 250 or more employees and an annual turn over of more than £25.9 million will pay a notification fee of £500. The £35 fee will continue to apply to other organisations. Failure to notify is a criminal offence under the DPA.

A further obligation which stems from the first principle is the requirement to provide certain information to individuals whose data is being processed.

1.5 Jurisdictional Scope

The DPA applies to a data controller established in the UK (and processing data in the context of that establishment); or a data controller established neither in the UK nor in any other EEA State but which uses equipment in the United Kingdom for processing the data otherwise than for the purpose of transit through the United Kingdom.

---

3   These are discussed in more detail at section 1.6 below.

Two terms used within the definition have required further interpretation; 'established' and 'use'. Firstly, the term 'established' has caused some debate at EU level as Recital 19 takes a sweeping approach and is purported to apply to any establishment that exercises effective and real activity through stable arrangements. Thankfully, the DPA adds clarification to this term by expressly stating that the definition includes individuals ordinarily resident in the UK; a body corporate registered in the UK; partnerships and incorporated associations formed under UK law and any offices, branches or agencies within the UK through which an activity is performed.

The second aspect of the definition that has caused controversy is the meaning of the term 'use'. The definition was considered by the Article 29 Data Protection Working Party after concerns were raised by a major internet search engine provider. The Working Party reiterated in 2008[4] that the Directive would apply to the use of cookies within the EEA even if the host website operated outside the territory. The Working Party had initially expressed this opinion in May 2002. In the Swedish case of Lindqvist[5] the ECJ was asked to consider, amongst other matters, whether Mrs Bodil Lindqvist's uploading of personal details about members of her parish onto a personal website amounted to a cross-border transfer of personal data. The court held that it was—the transfer occurred not at the point where personal data was uploaded onto a host site, but at the point it was accessed. The ECJ reasoned that the Directive had not envisaged that it would apply to the internet and in any case, there was no transfer until the data was actually accessed.

1.6 Key Compliance Requirements – the eight principles

One of the key requirements of the DPA is that data controllers process personal data in accordance with the eight principles listed in schedule 1 to the DPA.

First Principle: Personal data shall be processed fairly and lawfully and, in particular, shall not be processed unless: (a) at least one of the conditions in Schedule 2 is met, and (b) in the case of sensitive personal data, at least one of the conditions in Schedule 3 is also met.

The first principle is probably the most important—it requires that personal data be processed 'fairly and lawfully'. In practice this means a data controller's use of the relevant data must fall within a set of 'legitimate conditions' contained in schedules to the DPA including individual consent and the 'legitimate interest' condition.

Generally, 'fair' processing will not occur unless fair processing information is provided to, or is readily available to, the data subject in accordance with Schedule 1, Part II, paragraph 2(1) of the DPA. Such fair processing information includes the identity of the data controller, the purposes for which data is intended to be processed and any other information necessary for the processing to be fair (see paragraph 2, Part II, Schedule 1 of DPA).

An exception to the requirement to provide fair processing information can be found under Schedule 1, Part II, paragraph 2(1) of the DPA. Where the data controller has obtained the personal data from someone other than the data

---

4   WP 148: Opinion 1/2008 on data protection issues related to search engines.
5   *Bodi Lindqvist v Kammaraklagaren* (2003) C–101/01.

subject and it would impose 'disproportionate effort' to provide the fair processing information to the data subject, then the controller is relieved of the obligation to do so. Disproportionate effort is not defined under the DPA, and is a question of fact in each case. The Information Commissioner considers that the starting point is that data controllers are not generally exempt, and a number of factors need to be considered in assessing 'disproportionate effort', such as the length of time and cost involved to the data controller in providing the information. It should be noted that this exception does not absolve the data controller from its overriding duty to process data fairly.

The Information Commissioner has published a draft code of practice[6] which aims to help organisations draft effective privacy notices. In particular the draft guidance states that there is no need to 'tell people the obvious' and that there is no need to communicate fair processing information where the processing is something that the reasonable person is likely to anticipate and would consent to if asked.

'Fairness' also requires consideration of the circumstances in which the data was obtained and whether the person from whom the data was obtained was 'deceived or misled' as to the purposes for which it was to be processed (DPA, Schedule 1, Part II, paragraph 1(1)).

Schedule 1, Part II, paragraph 1(2) of the DPA states that data is to be treated as obtained fairly if it consists of information obtained from a person: (a) authorised by or under any enactment to supply it; or (b) required to supply it by or under any enactment or by any convention or other instrument imposing any international obligation on the UK. This provision may be of particular interest to certain public sector bodies.

The requirement of 'lawfulness' underscores the need to consider whether any processing violates any other applicable laws, for example, data disclosed in breach of a confidentiality obligation would not be disclosed lawfully and so would be in breach of the first principle. Compliance with the DPA will not render unlawful processing lawful.

In addition to the general fairness principle, certain pre-conditions have to be met as listed in Schedules 2 and 3. They refer respectively to further conditions that must be satisfied for the processing of personal data and additional conditions to be met if sensitive personal data[7] is to be processed. Personal data will generally only be considered as fairly processed if one of the following conditions is satisfied:

- an individual has consented to the processing;

- the processing is necessary to perform an activity for or involving the individual;

- the processing is necessary to comply with a legal obligation;

- the processing is necessary to protect the interest of the individual;

---

6   ICO Draft privacy notices code of practice issued 25 December 2008.
7   Sensitive personal data is information concerning an individual's race, racial or ethnic origin, political opinions, religious beliefs or other beliefs of a similar nature, trade union membership, physical or mental health condition, sexual life, commission or alleged commission of any offence, or proceedings for any offence committed or alleged to have been committed.

- the processing is necessary for a legitimate interest of the data controller or third party and it is not prejudicial to the individual;

- one of the other pre-conditions in Schedule 2 is satisfied.

Additional requirements are imposed by the DPA in respect of 'sensitive personal data'. The main differences are that a higher level of consent is required in respect of sensitive personal data and the 'legitimate interest' condition for processing does not apply to sensitive personal data.

Second Principle: Personal data shall be obtained only for one or more specified and lawful purposes, and shall not be further processed in any manner incompatible with that purpose or those purposes.

The 'specified and lawful purpose' refers to the purposes specified in the fair processing information provided to the individual or the purposes specified in the data controllers' notification with the ICO.

If a data controller wants to process personal data for a further purpose which is incompatible with the 'specified and lawful purpose' then the consent of the individual will be required. It is not sufficient to simply notify the individual of the further purposes or retrospectively amend the ICO notification for which the personal data is being processed. ICO guidance[8] makes it clear that the matter of compatibility is strictly regarded by the ICO and that when considering compatibility the ICO will take into account the purpose or purposes for which personal data is intended to be processed. An example of breach of the second principle came in 2006, when the SWIFT organisation, the worldwide financial messaging service which facilities international money transfers, was criticised by the Article 29 Working Party[9]. SWIFT shared customer data with the US government following subpoenas issued after the 9/11 attacks. The Working Party referred to the fact that the personal data had been collected by other financial institutions only for the purposes of processing client payment orders and subsequently by SWIFT for the purposes of executing its SWIFTNet FIN service. The Working Party concluded that SWIFT's continued processing of personal data, knowing the large scale of US Treasury Subpoenas, was a further purpose (ie for alleged-terrorism investigations) not compatible with the original specified commercial purpose for which the data was collected.

The second principle is often relevant in practice where a business expands into a new field and wishes to cross sell the new product or service to its existing customer base. This was considered in the *British Gas*[10] case where the company admitted that in sending its existing gas customers direct marketing relating to its new electricity business and non gas related services, it had breached the second principle. This processing of customer data between its supply and marketing databases was other than for a registered purpose and British Gas also admitted that the use of the data breached the third principle (see below). A similar situation occurred in *Midlands Electricity Plc v The Data Protection Registrar*[11], where the inclusion with

---

8   Legal Guidance on the Data Protection Act 1998 issued by the Information Commissioner and available at www.ico.gov.uk.
9   Opinion 10/2006 on the processing of personal data by the Society for Worldwide Interbank Financial Telecommunication (SWIFT).
10   *British Gas Trading Limited v The Data Protection Registrar* (24 March 1998).
11   7 May 1999.

customer bills of magazine advertising services and goods unrelated to the supply of electricity contravened the first principle as unfairly processing customer data.

Third Principle: Personal data shall be adequate, relevant and not excessive in relation to the purpose or purposes for which they are processed.

ICO guidance on this principle states that a *'wide definition of processing should be borne in mind when considering'* its application.

When processing personal data, the data controller should identify the minimum amount of information that is required to fulfil its purpose. If it is decided that more information is required about certain individuals then the data controller should only collect information in those specific cases.

It is also important to appreciate that there is a difference between holding information that **may be** useful in the future and information that has a foreseeable use despite the fact that the information may never be used. The former example would breach the obligations under the third principle whereas the latter would be allowed.

A data controller should consider conducting a 'privacy impact assessment' ('PIA') or similar exercise (which involves assessing the benefits that the data sharing might bring to society and individuals against the negative effects or likelihood of damages, distress or embarrassment to individuals) to ensure compliance. The PIA is a useful tool which should be considered as part of any project implementation; this is in contrast to audits which would be appropriate once a system was in place. In December 2007, the ICO published a Privacy Impact Assessment Handbook[12] to encourage private and public sector organisations to use PIAs which contains useful guidance on planning and conducting a PIA.

It is perfectly legitimate for parties receiving personal data from a third party to ask for unnecessary information to be removed before a transfer is undertaken to ensure that they are compliant. For example, it has been widely reported that when the National Audit Office requested the data that was lost by HMRC in 2007, it had requested that the fields of data that it was not interested in be removed. It has also been reported that HMRC decided that the expense of filtering out information (£5,000) was unjustified.

Fourth Principle: Personal data shall be accurate and, where necessary, kept up to date.

It is a general requirement of data privacy legislation that data should be accurate and where necessary, kept up to date and be adequate, relevant and not excessive for the legitimate purpose for which it is used. One approach to this is to have a privacy policy setting out guidance as to how individuals may update their data such as where their contact details change. Many organisations facilitate this through the use of an intranet or internet web sites.

Provisions contained within the DPA provide guidance on the fourth principle. As long as the data controller ensures that all reasonable steps have been taken

12   Published 11 December 2007 and available at www.ico.gov.uk.

to ensure the accuracy of the data and any inaccuracy highlighted by the data subject is indicated on the information collected, the principle will not be breached.

Data controllers should be aware though that it is not enough to conclude that the information is correct merely because it was obtained from the data subject or from a third party. This means it may be difficult to say whether a data controller has in fact taken all 'reasonable steps' to ensure accuracy—the ICO will judge each case on its particular facts.

Fifth Principle: Personal data processed for any purpose or purposes shall not be kept for longer than is necessary for that purpose or those purposes.

It is a general requirement of data privacy legislation that data should be kept no longer than necessary for the legitimate purposes for which it is being processed. Many other laws and regulations contain apparently conflicting requirements for data to be retained for specific periods. Data retention requirements imposed in the telecommunications sector are discussed at section 6 below.

One approach to achieving compliance with this obligation might be to implement a policy to review data on a regular basis according to a set of standards to decide whether it should be retained. The policy should also ensure that records which are being disposed of are securely and effectively destroyed. Such policies often establish a general long stop date beyond which data should be deleted unless there is a statutory requirement to retain the data (eg tax records).

Any policy issued by a data controller will need to take into account other regulatory and legal requirements impacting the retention period for particular categories of data where retention is likely to be an issue, such as in the case of HR and company records. Any policy should ensure that these are reviewed on a regular basis by HR staff.

Sixth Principle: Personal data shall be processed in accordance with the rights of data subjects under this Act.

This principle relates to contraventions of rights granted under sections 7, 10, 11 and 12 of the DPA. These are discussed at section 3 below.

Seventh Principle: Appropriate technical and organisational measures shall be taken against unauthorised or unlawful processing of personal data and against accidental loss or destruction of, or damage to, personal data.

European data protection legislation requires that, where a data controller engages a third party to carry out processing on its behalf, the data controller must ensure the processing is carried out under a contract containing particular provisions. This should be relatively straightforward in respect of most new agreements to be entered into as most service providers are becoming accustomed to being asked to sign up to such provisions. A broader issue is what to do with existing agreements that do not contain the appropriate provisions, of which there are likely to be many.

The DPA requires that if a third party is processing personal data on behalf of a data controller then that controller must choose a data processor that provides sufficient guarantees in respect of the technical and organisational security measures governing the processing as detailed above. As noted above, it follows that stringent security requirements will also be required where a data controller is to pass information to another data controller.

Further details of the security obligations imposed by this principle may be found at Part II of Schedule 1. In particular it is important to note that the DPA states that 'having regard to the state of technical development and the cost of implementing any measures' the security measures must ensure a level of security appropriate to:

- the harm that might result from such unauthorised or unlawful processing or accidental loss, destruction or damage as mentioned in the seventh principle; and

- the nature of the data to be protected.

The ICO has published a good practice note[13] aimed at advising data controllers which security measures must be put in place to ensure compliance with the seventh principle. A further source of guidance is contained in the numerous reports which have been issued in the aftermath of recent security reviews conducted by the UK government following high profile security losses[14]. These reports, and subsequent comments by the ICO, clarify that personal data contained on moveable media such as USB sticks and laptops should be encrypted so that data cannot be accessed if the device is lost. The benchmark for compliance is based around the international standard for information security management ISO 27001/27002. Another recent development has been the BSI standard which was published in June 2009[15].

Eighth Principle: Personal data shall not be transferred to a country or territory outside the EEA unless that country or territory ensures an adequate level of protection for the rights and freedoms of data subjects in relation to the processing of personal data.

This is a major topic which is addressed in section 2 of this chapter: International Data Transfer.

2   International Data Transfer

2.1   General Restrictions

The eighth data protection principle, described above, derives from Article 25(1) of the EC Data Protection Directive, which requires EU member states to provide that the transfer to a third country of personal data which is undergoing processing, or is intended for processing after transfer, may take place only if the third country in question ensures an adequate level of protection.

---

13 'Data Protection Good Practice Note—Security of personal information' published 22 November 2007.
14 The Poynter Report, the IPCC report, and the Walport data sharing report and the O'Donnell report, available at www.ico.gov.uk.
15 BSI Standard BS 10012:2009 available at www.bsigroup.com. See Section 8 below.

Guidance issued by the UK Information Commissioner[15a] clarifies that the eighth data protection principle does not apply to transfers of personal data which are 'mere transit'. This is a very helpful exclusion for the telecommunications industry as it recognises that packets of personal data may be directed through routers and other switches in a third country on its journey from one country to another as a natural result of the physical arrangement of international telecommunications infrastructure. So for example, where personal data being transferred from the UK to France passes through a switch or router in a non-EEA country, this will be 'mere transit' and will not be within the scope of the eighth principle.

The ICO has published a legal guidance document on the eighth data protection principle and international data transfer to assist data controllers in achieving this aim. The latest version of the Guidance[16] was published in December 2008 and takes into account a number of developments within the EU and at domestic level.

The Guidance sets out the ICO's recommended approach to assessing 'adequacy' for the purpose of transferring data outside of the EEA.

The European Commission is entitled to declare that the laws of a third country provide 'adequacy'. At the time of writing, positive findings of adequacy have been made in respect of Argentina, Canada, Guernsey, Isle of Man, Switzerland and Jersey; the 'white listed countries'. The situation regarding adequacy of protection in the USA deserves special consideration. The EEA entered into a special scheme with the USA in November 2000, known as the Safe Harbor scheme. Safe Harbor contains a set of principles similar to those in the DPA which govern specific transfers to the USA. For further discussion on Safe Harbor, please refer to section 2.2 of this chapter.

If a country has not been deemed as providing an adequate level of protection, then it is the duty of the data controller to assess adequacy, bearing in mind the objectives of the Directive and the DPA. In measuring adequacy, the data controller should have particular regard to Article 25(2) in the Directive and Schedule 1, Part II paragraph 13 in the DPA. Both provisions set out criteria for adequacy which the controller must satisfy before a transfer to a third country takes place. The ICO helpfully splits the criteria into two categories: 'the general adequacy criteria' and the 'legal adequacy criteria'. It is envisaged by the ICO that the general adequacy criteria should be applied first as they are easier to determine and could save cost and time if found that transfer is low risk. Practically, the data controller should consider the following factors first:

- the nature of the personal data;

- the purpose or purposes of the proposed transfer;

- the period during which the data will be processed;

- the security measures that will be taken in the third country;

- the country of origin of the personal data; and

- the country of final destination.

15a ICO Guidance: The eighth data protection principle and international data transfers.

16 ICO Guidance: The eighth data protection principle and international data transfers version 3.0 available at www.ico.gov.uk.

If after assessment of the general adequacy criteria it is found that the proposed transfer is high risk, then a thorough investigation of the legal adequacy should be conducted. Issues to consider under this limb will include:

■ the law in force in the third country;

■ the international obligations in the third country; and

■ any relevant codes of conduct or other rules that are enforceable in that country's territory.

Clearly, this sort of investigation will require highly specialised knowledge and data controllers should consider the commercial viability of their project before concluding that the transfer will be deemed adequate by regulators.

If a data controller feels that the international transfer would not meet the adequacy test, there are still further options available to him. These are discussed further in section 2.2 of this chapter.

2.2 Options for Transfer

There are a number of options open to a data controller seeking to transfer personal data outside the EEA in compliance with the DPA and these are summarised briefly in the following sections.

It is however important to bear in mind that addressing the international transfer problem is only one aspect of data protection compliance and other obligations in relation to the processing of personal data must also be complied with, in particular the requirement to ensure that the transfer is 'fair and lawful' as required by the first principle.

2.2.1 Consent

Transfers are permitted where the data controller has obtained consent to that transfer from the individual concerned. Although it may appear an attractive option it has been described by the Article 29 Working Party as a 'false good approach'[17] and that it may prove 'simple at first glance but in reality complex and cumbersome'.

Guidance[18] issued by the ICO requires that valid consent must be freely given, specific, informed and can be withdrawn at any time. This Guidance notes that it may not be possible to obtain 'freely given' consent in the employment environment. The Information Commissioner's view is supported by guidance from the Article 29 Working Party[19] where it states that:

---

17 Working document on a common interpretation of Article 26(1) of Directive 95/46/EC WP 114.

18 *Employment Practices Code ('Code') and the Supplementary Guidance ('Supplementary Guidance')* available at www.informationcommissioner.gov.uk.

19 The Article 29 Working Party was set up under the Directive to represent the data protection regulators from the different jurisdictions in forming a common approach to the implementation of the Directive.

'*specific difficulties might occur to qualify a data subject's consent as freely given in an employment context, due to the relationship of subordination between employer and employee*'.[20]

Employee consent is a particular problem in respect of existing workers, although it may be less of a constraint in relation to the recruitment process, provided the applicant is aware of exactly what he or she is consenting to.[21]

2.2.2 Necessary for the purposes of a contract

Transfers of personal data are also permitted where the transfer is necessary for a contract between the data controller and data subject or for a contract between the data controller and third party which is in the interest of the data subject.[22]

In practice this is quite a limited exception with a narrow interpretation of what is considered 'necessary'. Guidance issued by the Article 29 Working Party[23] gives the example that the outsourcing of payroll to a country that did not offer adequate protection may be beneficial and cost effective to a company but would not be considered necessary for the fulfilment of an individual's employment contract.

2.2.3 Model Clauses

One option commonly used in cross-border transfers of personal data is the incorporation of 'Model Clauses' into data transfer agreements. Essentially, the Model Clauses are a set of standardised clauses that have gained prior approval by the European Commission.

The European Commission has approved three sets of Model Clauses which it recognises as providing an 'adequate level of protection' when used to govern a transfer of data. Two sets ('Standard Clauses')[24] were approved in 2001; one set deals with transfers to controllers and the other deals with transfers to processors. An alternative to the Standard Clauses for transfers to controllers, based on a draft provided by the International Chamber of Commerce, ('ICC Clauses') were approved in 2004[25]. A fourth set, an alternative for transfers to processors, is due to be published in the near future but at the time of writing, the terms have not been finalised. The Article 29 Working Party has published its views on a draft set of these clauses[26]. This fourth set of clauses will be of particular importance to the outsourcing industry as the stated intention is to provide a more practicable solution to the way in which sub-processors are dealt with.

---

20 WP114, *Working document on a common interpretation of Article 26(1) of Directive 95/46/EC*, adopted 25 November 2005.
21 This is on the grounds that individuals in the open job market will usually have a free choice whether or not to apply for a particular job.
22 Article 26 (1)(b) and Article 26 (1)(c), Directive 95/46/EC.
23 WP114, *Working document on a common interpretation of Article 26(1) of Directive 95/46/EC*, adopted 25 November 2005, page 13 and page 14.
24 Commission Decision of 27 December 2001 on standard contractual clauses for the transfer of personal data to processors established in third countries, under the EU Data Protection Directive, reference C(2001)4540, and Commission Decision of 15 June 2001 on standard contractual clauses for the transfer of personal data to third countries, under Directive 95/46/EC, reference C(2001) 1539.
25 Commission Decision of 27 December 2004 reference C(2004) 5271.
26 Article 29 Working Party opinion 3/2009 on draft Commission Decision WP161.

The ICC Clauses are considered to be more 'business friendly' than the Standard Clauses. Although they are, in substance, very similar to the Standard Clauses, there are some significant differences between the ICC Clauses and the Standard Clauses, primarily in relation to liability. It is also important to note that the ICC Clauses do not cover transfers between data controllers and data processors.

In approving the Model Clauses the European Commission placed a requirement on national privacy regulators to approve transfers made pursuant to the Model Clauses where the exact form of wording which appears in the Model Clauses is used. Despite this some national privacy regulators[27] nevertheless insist on a requirement for formal approval of transfers according to the exact form of the Model Clauses even though they are obliged to approve such transfers in accordance with European law.

2.2.4   Use of an Intra-Group Agreement

The Model Clauses may be incorporated into an Intra-Group Agreement ('IGA') to enable the transfer of personal data to entities within a group. The attraction of an IGA is that it avoids the need to enter a large number of point to point agreements.

The possible structure of a typical IGA can be shown diagrammatically as follows:

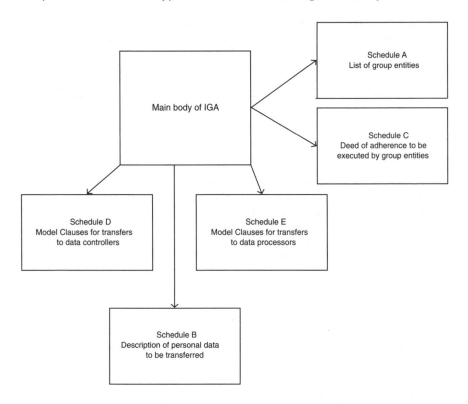

27   Including the data protection regulators in Spain and Portugal.

There are a number of ways in which the IGA can be structured to bind all the entities in the group. The preferred approach will depend on various factors including the number of entities, the regional governance arrangements, the way in which risk and liability is apportioned through the group and the way in which data is transferred and used within the organisation.

IGAs can be complex documents, at least in part because the Model Clauses themselves are complex. However, when used properly they can greatly reduce the administrative burden an organisation would otherwise be faced with implementing a large number of 'point to point' Model Clause contracts between its group members.

Any business considering the adoption of an IGA should also understand that approval by some national regulators may be required.

2.2.5   Safe Harbor

GENERAL

The Safe Harbor programme was implemented in 2000 by agreement between the European Commission and the US Department of Commerce. The purpose of Safe Harbor is to provide a method to comply with the restrictions on exporting personal data outside the European Economic Area ('EEA') to the US. European Union Member States are required to recognise transfers under Safe Harbor as compliant with European data privacy legislation. US personal data importers who certify compliance with the Safe Harbor Principles are deemed '*adequate*', so that data may lawfully be sent from the EEA to importers in the US.

THE CERTIFICATION PROCESS

US entities which are subject to the jurisdiction of the FTC or the Department of Transport may participate in the US Safe Harbor programme. In order to participate, and be included on the 'Safe Harbor list' maintained by the US Department of Commerce, the company importing the data into the US must agree to comply with the seven Safe Harbor Principles and the 15 FAQs which together comprise the Safe Harbor Framework[28].

2.2.6   Binding Corporate Rules ('BCR')

GENERAL

The BCR approach involves all entities within a group organisation adopting a set of binding common data privacy standards that ensure an adequate level of protection is afforded to personal data and that data subjects' rights will not be prejudiced as a result of transfers made to countries outside the EEA. The intention is to provide international businesses with a pragmatic, feasible solution to international data transfers and avoid the need for the use of individual Model Clauses contracts.

---

28   These documents are available at www.export.gov/safeHarbor and impose standards broadly equivalent to the standards of data protection contained in the EU Data Protection Directive.

The BCR approach requires organisations to provide a degree of supporting information including a description of the data flows and compliance measures within the organisation along with details of how the BCRs will be enforceable against the group members and by individual data subjects.

Although the benefits of the BCR approach are attractive to most businesses, in practice only a relatively small number of companies to date have the necessary approvals, including General Electric, Phillips, Accenture and, most recently, Hyatt. The primary reason for this is that the approval process has been seen as too costly and cumbersome for most corporate institutions. There has also been a feeling that organisations might attract unwelcome attention from regulators if they make a BCR application as the process has been quite high profile. However, recent developments in the manner in which approval is sought may mean that an increasing number of organisations decide that the BCR approach is a sensible approach for them to take.

STRUCTURE

Various guidance documents published by European data protection regulators set out at a high level the requirements for a set of BCRs.[29]

The BCR approach requires the disclosure of details of the group's processing activities and compliance measures to a greater level of detail than will be required in an application for approval of an IGA. However, organisations which have adopted an IGA have found that this can used as part of the compliance package which forms the BCR application.

In June 2008, the Article 29 Working Party[30] ('WP') created a 'toolbox' for composing BCR. This toolbox consists of:

- a framework document that sets out an outline for the structure and content of BCR in line with EU data privacy requirements[31];

- a BCR checklist that sets out the necessary content of BCR and distinguishes that from information that must be provided in support of a BCR application[32];

- a set of FAQs which provide answers to questions that are commonly raised in practice[33].

---

29 For example: Article 29 Working Party Paper WP 74, Working Document on Transfers of personal data to third countries: Applying Article 26(2) of the EU Data Protection Directive to Binding Corporate Rules for International Data Transfers, 3 June 2003; and Article 29 Working Party Paper WP 108, Working Document Establishing a Model Checklist Application for Approval of Binding Corporate Rules, 14 April 2005. Article 29 Working Party Paper WP154, Working Document Setting up a Framework for the Structure of Binding Corporate. Article 29 Working Party Paper WP155, Working Document on Frequently Asked Questions (FAQs) related to Binding Corporate Rules.
30 A formal body set up under the Data Protection Directive consisting of representatives from each EU Member State Data Protection Authority. The Article 29 Working Party issues guidance on the interpretation and implementation of the Data Protection Directive.
31 Article 29 Working Party Paper WP154, Working Document Setting up a Framework for the Structure of Binding Corporate.
32 Article 29 Working Part Rules WP153 Working Document Setting up a table with the elements and principles to be found in Binding Corporate Rules.
33 Article 29 Working Party Paper WP155, Working Document on Frequently Asked Questions (FAQs) related to Binding Corporate Rules.

In order for a company's BCRs to be used for the transfer of personal data from EU Member States to countries that do not offer an adequate level of protection for personal data, approval of the BCRs must be given by each Member State from which personal data will be transferred. This is an onerous burden on any organisation wishing to implement BCR.

However, in October 2008 the Working Party announced the implementation of a 'mutual recognition procedure'. The mutual recognition procedure involves the nomination of a 'Lead Authority'. This is the data protection regulator of a Member State which will be responsible for progressing the BCR approval process. The Lead Authority will be nominated by the company developing the BCRs. The Lead Authority will have responsibility for reviewing all documents submitted, negotiating the BCR and finally approving the BCR. All other participating data protection authorities, that are part of the 'mutual recognition procedure', should accept the Lead Authority's findings. Presently, though, only Cyprus, France, Germany, Ireland, Italy, Latvia, Liechtenstein, Luxembourg, the Netherlands, Norway, Spain and the UK participate in mutual recognition procedure.

3 Personal Data and individual's rights

3.1 Introduction

Part II of the DPA confers a number of rights on individual data subjects, including:

- a right to access personal data;
- a right to rectify erroneous information;
- a right to prevent data processing which is likely to cause damage or distress;
- a right to prevent direct marketing; and
- a right to prevent automated decision making.

The right to access personal data is set out in Sections 7 to 9 of the DPA and entitles individuals to submit 'access requests' to data controllers, who must then confirm whether or not they are processing any personal data relevant to the individual and if they are, provide both a description of the processing being carried out and a copy of the relevant personal information.

The DPA recognises that information disclosure under the subject access regime may not be appropriate in all cases—for example where providing access to one individual would conflict with the expectation of privacy owed to another individual, or where disclosure would cause undue prejudice to the legitimate interests of the individual or those of a third party. Exemptions exist to cover many of these circumstances.

3.2   Responding to a Subject Access Request

Time limit for response

Once a data controller has established that a subject access request is legitimate, they will need to provide the data subject with access to relevant information within certain prescribed time limits (40 days in most cases).

Information to be provided

Section 7(1)(a) sets out the information that the data controller must provide to the data subject in response to a valid request, comprising of an explanation of:

- the personal data that is held,
- the purposes for which the data is processed;
- the likely recipients of the data;
- the sources of the data;
- where the data is used to make automated decisions about the individual, the logic involved in that decision taking; and
- a copy, in permanent form, of the information held (unless this would require disproportionate effort or would not be possible—in which case other means of access should be provided).

By way of example, if the data subject is an employee, this may involve the employer providing an explanation as to the categories of data retained in HR files, reference to the fact that disclosures are made to third party payroll providers and providing a hard copy of the electronic and paper records held within the HR department and by line managers regarding the employee.

3.3   Failure to comply

If the data controller fails to respond to a compliant subject access request the data subject may apply directly to the courts under Section 7(9) for an order requiring the data controller to comply with the request. If that failure additionally causes the individual 'damage and distress' then the individual may claim compensation for damage under Section 13 of the DPA. Note that:

- Data controllers will have a defence to Section 13 claims if they can show that they took such care in the circumstances as was reasonably required to comply with the relevant requirements of the DPA.
- The courts have also shown a general reluctance to grant orders under Section 7(9) where the request substantively relates to matters which are the subject of other legal processes and to order disclosure would distract from or circumvent the proper following of those processes.

An alternative and somewhat more straightforward route of complaint for an aggrieved data subject is to make a statutory request to the Information Commissioner to determine whether the data controller has effectively satisfied its responsibilities under Section 7. The Information Commissioner must consider all complaints received and may, if he decides to investigate the matter further, serve a formal information notice on the data controller requiring it to provide information regarding the way in which it handled the request and/or if he believes the DPA has not been complied with, serve an enforcement notice. Failure to comply with either an information or enforcement notice is a criminal offence under Section 47.

The Information Commissioner has not been shy of taking formal action against organisations in flagrant breach of the rules—for example, in 2006, Liverpool City Council was prosecuted for failing to comply with an information notice served following a complaint about the Council's inadequate response to a subject access request from a former employee. In that case, the District Judge fined the Council £300 and publicly condemned its processes as showing 'appalling breakdown of communication' and 'a clear lack of compliance' with the provisions of the DPA[34].

Whilst the Information Commissioner has been quick to tackle organisations that have failed to observe their obligations under the DPA, a more conciliatory message has been issued for those who make every effort to comply with the rules, but find it difficult to achieve full compliance in all cases. The Information Commissioner has for instance indicated that where compliance would cause an organisation 'disproportionate effort', he will have regard to all relevant factors before determining the approach he will take to enforcement.[35]

3.4 Meaning of Personal Data

There has been a great deal of controversy surrounding the definition of 'personal data' in recent years.

The first case to consider the matter was *Durant v Financial Services Authority*.[36] ('Durant')

The facts of *Durant* can be summarised as follows: D was a customer of Barclays Bank. He engaged in litigation with Barclays, which he lost. He then complained about Barclays to the Financial Services Authority (FSA). The FSA investigated the complaint but closed its file without telling D the outcome, asserting a statutory duty of confidentiality. D made two data subject access requests to the FSA seeking disclosure of personal data it held on him on both computerised and manual filing systems. The FSA provided D with some documents held on computerised filing systems, redacted to prevent disclosure of information to D which related to third parties. The FSA refused to provide any data to D which was held on manual files on the basis that it did not come within the definition of a 'relevant filing system'. On appeal to the Court of Appeal, D's appeal against the decision of the county court in favour of the FSA was dismissed.

The Court of Appeal focused on the requirement that data must 'relate to' an individual in order to constitute 'personal data' under the DPA. According to

---

34 See the Information Commissioner's Office Press Release entitled 'Liverpool City Council prosecuted for data protection offences', 14 December 2006.
35 See the Information Commissioner's Enforcement Statement, June 1999.
36 *Michael John Durant v Financial Services Authority* [2003] EWCA Civ 1746, Court of Appeal (Civil Division) decision of Lord Justices Auld, Mummery and Buxton.

the court, two factors are relevant in deciding whether the data 'relates to' an individual:

- whether the data is biographical in a significant sense, going beyond the recording of the mere involvement of the data subject which has no personal connotations; and

- whether the information has as its focus the data subject, rather than some other person with whom he may have been involved, or some transaction or event in which he may have figured or had an interest.

The court held that the mere mention of an identifiable individual within a pack of information will not itself constitute personal data: the information must relate to an aspect of the individual's personal privacy to be regarded as 'personal data'.

The effect of *Durant* is to suggest a narrower definition of the scope of information which is to be regarded as within the scope of the subject access rules than previously adopted by practitioners. As a result, for example, minutes of a business meeting which record individuals' attendance will not be accessible by those individuals; it is only where the discussion specifically refers to the individual's personal conduct or another matter specific to the individual's circumstances that disclosure can be anticipated.

Whilst this approach has been affirmed by the courts in subsequent cases including *Smith v Lloyds TSB Bank Plc*[37], there are genuine concerns that UK case law in this area is in conflict with the (wider) concept of 'personal data' as envisaged under the supervening EC Data Protection Directive and explained in further detail in a recent opinion presented by the Article 29 Working Party[38].

Opinion 4/2007 of the Working Party was adopted on 20 June 2007 to offer guidance on the concept of personal data following uncertainty and diversity in practice among Member States. The aim was to come to common understanding of personal data to ensure data protection rules are enforced in a uniform manner. It essentially analyses the four key elements of the definition contained in the Directive:

- 'any information'—the Working Party considered that this is to be interpreted widely and covers both objective and subjective information, whether or not it is accurate;

- 'relating to'—in general terms information will be judged to relate to an individual if it is *about* an individual. This is often obvious but there are cases where it is not self-evident, such as where the information primarily relates to an object. One of three elements should be present: content (it is about the individual); purpose (it is used to evaluate or in some way affect the treatment of the individual); and result (it may have an impact on the individual or their rights or interests);

- 'an identified or identifiable'—the Working Party felt that this is most often satisfied by a reference by name but would include any pieces of information which, either separately or taken together, could identify a person;

---

37 [2005] EWHC 246 Ch.
38 See Opinion 4/2007 of the Article 29 Working Party on the concept of personal data (WP 136).

- 'natural person'—the Working Party reiterated that the definition covers living human beings and is not restricted to nationals or residents of any particular country.

In forming their opinions, the Working Party seemed to be implicitly challenging the decision of the UK courts and the UK's implementation of the Directive.

The ICO is acutely aware of these conflicting positions and in response published updated guidance in an attempt to reconcile the seemingly conflicting approaches taken by the UK court in *Durant* and the subsequent Article 29 Working Party paper described above. The updated guidance took the form of a Good Practice Note which provides a useful decision chart to follow when considering whether information is personal data for the purposes of the DPA[39]. However, although the question of what information comprises personal data is clear in the overwhelmingly majority of cases, inconsistencies and uncertainty remains.

The UK courts sent a clear signal in recent cases including *Ezsias v Welsh Ministers*[40] that they will continue to follow Durant. Although *Ezsias* was decided after the later ICO Good Practice Note, the note was not referred to in the judgment and the principles in Durant were applied by the High Court. In choosing to ignore the ICO guidance the Court gave a sharp reminder that, unlike the decisions of the higher court (such as the Court of Appeal in Durant), such ICO guidance is not binding on the Courts. However, data controllers should not take this as a signal that the ICO guidance should be ignored as it remains the benchmark for compliance with the DPA on this and wider issues of DPA compliance.

3.5 Exemptions to disclosure

In some circumstances, a data controller will not be under any obligations to respond to a subject access request. There are a number of limited restrictions and exemptions to these rights of access including where to provide a copy (not the information itself) would involve disproportionate effort, where the disclosure cannot be made without disclosing information relating to other individuals, or where the disclosure would reveal information likely to prejudice ongoing negotiations. Legally privileged information is also exempt and there are other exemptions including those relating to the disclosure of management forecasting and planning information which would be likely to prejudice the business. Section 29 of the DPA makes personal data processed for the purposes of crime prevention or detection, or the assessment or collection of any tax exempt from disclosure under section 7 of the Act.

4 Regulation of Investigatory Powers Act 2000

4.1 Introduction

In February 2000, the UK government proposed a bill intended to allow law agencies to monitor terrorists, smugglers and other criminals using sophisticated

---

39   See Data Protection Technical Guidance – Determining what is personal data (29 August 2007).
40   *Ezsias v Welsh Ministers 2007ALL ER (D) 65*

technology systems. The bill was justified as a clarification of existing powers that police had to engage in covert surveillance. The first proposal presented to both the houses was considered unduly wide and fears were raised of misuse and potential infringement of civil liability. A revised version was proposed in the face of sustained opposition and the monitoring of communications is now generally governed by the Regulation of Investigatory Powers Act 2000 ('RIPA'). It deals with both criminal and civil liabilities for monitoring telephone calls and other communications in certain circumstances. In April 2009, the government announced a consultation and review of RIPA in light of concerns regarding the extent to which local authorities are using surveillance powers that are only intended to be used to fight terrorism following the events of 11 September 2001. Concerns have been raised as to the necessity and proportionality of the widespread use of these powers by local authorities. The consultation sought views on which local authorities should be able to use RIPA powers and for what purposes. Comments have also been invited on the level of authorisation that should be required for local authorities to use RIPA powers and whether this level needs escalating.

4.2 Definitions

Similar to the DPA, RIPA uses a number of technical definitions that are worth setting out:

- 'telecommunication system' means any system for communication by electrical or electro-magnetic energy[41], and so will include any system of telephones;

- 'private telecommunication system' means any telecommunication system (other than a public telecommunications system) which is attached to a public telecommunication system[42], and so is likely to include any internal telephone system which is or can be connected to the general public telephone system;

- 'public telecommunication system' means any telecommunication system which is available to a substantial section of the public in the UK[43];

- 'system controller' means any person with a right to control the operation or use of a private telecommunication system;

- an 'interception' of a communication includes the monitoring of that communication by means of the telecommunication system so as to make all or part of the communication available to a person other than the sender or recipient (whether at the time or storing it for later)[44]. It does not include the recording of any telephone conversation by either of the parties on the call, but would include recording or listening in by a third party using that telecommunication system.

4.3 Statutory Provisions

It is a criminal offence under RIPA to intentionally and without 'lawful authority' intercept in the UK any communication transmitted by means of a public postal service or a public telecommunications system[45].

---

41 Section 2(1), RIPA.
42 Section 2(1), RIPA.
43 Section 2(1), RIPA.
44 Sections 2(2), (7) and (8), RIPA.
45 Section 1(1), RIPA.

It is also an offence under RIPA to intentionally and without 'lawful authority' intercept any communication transmitted by means of a private telecommunications system[46], unless the interception is made by, or with the express or implied consent of, the system controller[47].

However, any interception of such a communication transmitted by means of a private telecommunications system by or with the consent of the system controller, without 'lawful authority', gives rise to a civil liability actionable by the sender or recipient of the communication[48].

For the purposes of the provisions described above, 'lawful authority' can take a number of forms, including where both the sender and the recipient of the communication have consented to the interception (or where the person making the interception has reasonable grounds for believing that such consents have been provided).[49] Otherwise, it may be deemed lawful by reason of the Telecommunications (Lawful Business Practice) (Interception of Communications) Regulations 2000 (SI 2000/2699) (the 'LBP Regulations') discussed further below.

4.4 Lawful Business Practice Regulations

The LBP Regulations give several other circumstances where 'lawful authority' is deemed. These include interception:

- for the purpose of establishing the existence of facts[50];

- for the purpose of ascertaining compliance with regulatory or self-regulatory practices or procedures[51];

- for the purpose of checking that employees are meeting certain standards when using the system in the course of their duties[52]; and

- for the purpose of preventing or detecting crime[53].

Interception for these purposes is only permitted where (i) the intercepted communication is relevant to a business[54], (i.e. that it is a communication by means of which a transaction is entered into in the course of the business, or which otherwise relates to or takes place in the course of carrying on that business); (ii) the telecommunication system in question was provided for use wholly or partly in connection with that business; and (iii) the system controller has made all reasonable efforts to inform every person who may use the telecommunication system in question that communications transmitted by it may be intercepted[55].

---

46 Section 1(2), RIPA.
47 Section 1(6), RIPA.
48 Section 1(3), RIPA.
49 Section 3(1), RIPA.
50 Regulation 3(1)(a)(i)(aa), LBP Regulations.
51 Regulation 3(1)(a)(i)(bb), LBP Regulations.
52 Regulation 3(1)(a)(i)(cc), LBP Regulations.
53 Regulation 3(1)(a)(iii), LBP Regulations.
54 Regulation 3(2), LBP Regulations.
55 Regulation 3(2), LBP Regulations.

Practical Issue: Recording Calls

In the case of incoming calls, provided a business is only 'intercepting' (recording) for one of the purposes permitted by the LBP regulations, it will not infringe RIPA provided it uses reasonable efforts to inform every person who may 'use' the system of the interception. It is clear that employees 'use' the system and so must be informed but it is less clear as to whether third party callers (or senders of email) would also need to be informed. The ICO acknowledges in guidance[56] that it 'is hard to see how consent can readily be obtained from external senders of email…' and so presumably acknowledges that such consent cannot be always obtained by using 'reasonable efforts' as required by RIPA. In any event, it would be unlawful to intercept calls for any other purposes (such as marketing or market research) without consent.

Even if 'lawful authority' is obtained under the LBP Regulations, the business must still comply with the requirements of the DPA. As set out at section 1.6 above, under the DPA personal data (namely data which can identify living individuals, including a call recording where the caller can be identified) may only be 'processed' (which includes using, storing and accessing) in accordance with the 8 data protection principles. Although businesses must comply with all of the principles if a recording contains personal data, the first, second and fifth principles are arguably most relevant to call monitoring and recording.

First principle—fairly and lawfully

If a business relies on consent (via a recorded message or telephone operatives' greetings, including a warning) to obtain lawful authority under RIPA to intercept incoming calls, then that consent will also meet the requirements of the first principle.

There are, however, other ways to meet the requirements of the first principle. In this context the most relevant ways are likely to be that the recording or monitoring is:

- done by a business for its 'legitimate interests';

- 'necessary for the performance of a contract to which the… [caller] is a party, or for the taking of steps at the… [caller's] request … with a view to entering into a contract'; or

- (where a regulatory authority requires call recording) 'necessary for compliance with any legal obligation to which … [the business] is subject, other than an obligation imposed by contract'.

Second principle—specified and lawful purposes

The data subject/caller must be told at the outset what uses will be made of his/her data. Where calls are recorded for marketing or market research purposes (ie outside the purposes contemplated by the regulations), the data subject's consent is required and a business cannot justify processing for these purposes on the basis of the exceptions to the first principle outlined above.

---

56   ICO: The Employment Practices Code Supplementary Guidance available at www.ico.gov.uk.

Fifth principle—not kept longer than necessary

The purpose for which a recording is made will affect how long it would be reasonable for it to be retained. There would be little justification in keeping recordings made to verify what was said for longer than six years (or six years following termination of any related contracts), as after such time the Limitations Act 1980 prevents the bringing of legal proceedings. Where recordings are made for training or quality monitoring purposes, a business would only be justified in keeping such recordings for a much shorter time. If, however, a business wanted to keep a recording for longer, one option would be to 'anonymise' the message (by deleting those parts of it that identify individuals) so that it contains no personal data. The recording could then be kept indefinitely.

Best practice as regards employees and third party callers

Part 3 of the Information Commissioner's Employment Practices Code (issued in June 2005 under s 51 DPA) gives guidelines on monitoring employees and, to some extent, third parties in accordance with the DPA. Although the code is not legally binding, the commissioner will likely reference it in any enforcement action.

The code states and advises that:

- employees have a legitimate expectation to privacy for their personal lives and are entitled to a degree of privacy in the work environment;

- employees should be made aware of the nature, extent and reasons for monitoring, preferably by use of prominent signs in areas subject to monitoring. Telling employees that from time to time they may be subject to audio monitoring is not sufficient. The code suggests a rule of thumb for testing the fairness of monitoring—whether, at the point at which monitoring occurs, employees would be aware it was taking place;

- where possible, monitoring should be targeted at areas of particular risk or confined to situations where employees' expectations of privacy will be low;

- covert monitoring is only justified in exceptional circumstances;

- where the monitoring of incoming calls goes beyond listening in on such calls without recording them, and therefore involves the processing of personal data, incoming callers should be told monitoring is taking place and the reasons for it unless this is obvious;

- where reasonably practical, this information should be provided by way of recorded messages on telephone systems, or by operatives' greetings including a warning that calls may be recorded and the reasons for this. There is no prescribed wording to use, but a good starting point would be: 'Please note that calls may be recorded for training, quality monitoring or evidential purposes'.

Finally, it is good practice for a business to inform employees of any monitoring, and how employees' own personal data will be used, by incorporating a data protection policy directly (or by reference) in its employment contracts.

5 Privacy and Electronic Communications Regulations 2003

5.1 Introduction

The Privacy and Electronic Communications (EC Directive) Regulations 2003 ('PEC Regulations') cover a wide range of activities. The Regulations, which came into force on 11 December 2003, implement EC Directive 2002/58/EC, and were designed to be more 'technology neutral' than the Telecommunications (Data protection and Privacy) Regulations 1999 that they replaced. This was so that any developments in electronic communications technology can be addressed within the framework of the 2003 Regulations.

The PEC Regulations require that providers of a public electronic communications service take appropriate technical and organisational measures to safeguard the security of that service[57] and require that subscribers must be informed if the security of the service is compromised[58].

The PEC Regulations also address, amongst other things, the use of location and traffic data, the use of cookies, direct marketing by electronic means and caller line identification services. These are addressed in the following sections.

5.2 Location Data

In recent years there has been a massive increase in the use of location data[59].This is due to the expansion in the use and availability of satellite GPS systems and also of the widespread use of mobile phones, through which users can potentially be located. Companies have been quick to spot revenue-generating uses of such data. The PEC Regulations regulate the use of location data at Regulation 14.

The first services making use of location data which were offered to individuals gave information relevant to that individual's position such as the nearest train station, or a particular show being staged at a nearby theatre. More recently, services are being offered which locate individuals at the request of third parties, most commonly parents wishing to locate children or employers wanting to locate employees.

Regulation 14 of the PEC Regulations imposes certain restrictions on the use of location data. The DPA is also relevant to the processing of location data as such data will invariably relate to identified or identifiable individuals

Regulation 14 provides that location data relating to a user or subscriber of a public electronic communications network or a public electronic communications service may only be processed where necessary for the provision of a value added service, with the consent of that user or subscriber.

---

57 Privacy and Electronic Communications Regulations 2003 Regulation 5(1).
58 Privacy and Electronic Communications Regulations 2003 Regulation 5(2).
59 Location data is defined as 'any data processed in an electronic communications network indicating the geographical position of the terminal equipment of a user of a public electronic communications service, including data relating to the latitude, longitude or altitude of the terminal equipment; the direction of travel of the user; or the time the location information was recorded.'

Before obtaining this consent the PEC Regulations require that the individuals who are the subjects of location data which is processed are informed of the following:

- the type of location data being processed;
- the purposes of the processing;
- the duration of the processing; and
- whether the data will be transferred to a third party for the provision of the value-added service.

Individuals must also be told that they have the right to:

- access and rectify their data;
- withdraw their consent at any time; and
- refuse to allow their data to be processed temporarily.

In particular, the ICO has issued guidance stating that a user or subscriber should be offered the opportunity to withdraw consent on the occasion of each connection to the network or on each transmission of a communication[60].

According to the Article 29 Working Party[61], the obligation to inform individuals falls on the party which collects the location data for processing. This will generally be the provider of the value-added service, but in those cases where the service provider does not have direct contact with the data subject, the electronic communications operator should provide the information as this is the party which has ultimate responsibility for complying with the PEC Regulations.

Under the provisions of the PEC Regulations, any service provider supplying a value-added service which involves the processing of location data must obtain the consent of the users or subscribers before such processing takes place. In addition, the users and subscribers must be informed of the terms of the processing.

The PEC Regulations do not specify how consent should be obtained. However, the Information Commissioner has issued guidance[62] stating that in order to obtain valid informed consent, the subscriber or user should be given enough clear information for them to have a broad appreciation of how the data is going to be used and the consequences of consenting to such use. This means that the service provider will not be able to rely on a blanket 'catch all' statement but will need to obtain specific informed consent for each value added service provided.

The guidance goes on to consider the issue of which party should obtain consent in the situation where a value-added service is provided by a communications provider in conjunction with a third party. The recommendation is that the party who will be seen to be responsible for providing the service is the one which should obtain consent, the key message being that the way in which the service is provided should be consistent with the expectations of the subscriber or user.

---

60   ICO Guidance on the Privacy and Electronic Communications (EC) Directive Regulations 2003. Part 2. Version 3.4 dated 30 November 2006.
61   Article 29 Working Party opinion on the use of location data with a view to providing value added services WP 115 25 November 2005.
62   ICO Guidance on the Privacy and Electronic Communications (EC) Directive Regulations 2003. Part 2. Version 3.4 dated 30 November 2006.

The person from whom the consent is required is, in the case of a service offered to private individuals, the person to whom the data refer. However, where there is a corporate subscriber, the person who makes decisions on behalf of the company is likely to be able to consent for that subscriber.

5.3 Traffic data

Regulations 7 and 8 of the PEC Regulations regulate the processing of traffic data which is defined as:

> *'any data processed for the purpose of the conveyance of a communication on an electronic communications network or for the billing in respect of that communication'*[63].

The Regulations contain limitations on the length of time for which traffic data can be retained and require that traffic data must be deleted or anonymised, when it is no longer necessary to transmit a communication provided that data required to calculate the subscriber's bill can be retained during the period in which the bill may lawfully be challenged. This would, under English law[64], suggest a period of six years, although the ICO[65] has indicated that this should not permit the wholesale retention of traffic data for six years and should only allow retention of data for longer than the normal billing cycle where a challenge is made.

Regulation 7 provides that traffic data may also be processed to provide value added services to the subscriber or the user; or to market the service provider's own electronic communication services. In each case the processing is only permitted where the subscriber or user to whom the traffic data relates has given his consent to value added services.

Value added service is defined as

> *'... any service which requires the processing of traffic data or location data beyond that which is necessary for the transmission of a communication or the billing in respect of that communication.'*

Only a communication services provider, or a person acting under their authority, can carry out the processing.

Regulation 8 provides that the public communications provider must provide the subscriber or user to whom the data relates with information regarding the types of traffic data which are to be processed and the duration of such processing. Regulation 8 also provides that the activities which can be carried out without consent are limited to (a) the management of billing or traffic; (b) customer enquiries; and (c) the prevention or detection of fraud.

---

63 'Public communications provider' is defined as 'a provider of a public electronic communications network or a public electronic communications service'.
64 Limitations Act 1980.
65 Paragraph 3 ICO Guidance on the Privacy and Electronic Communications (EC) Directive Regulations 2003. Part 2. Version 3.4 dated 30 November 2006.

5.4 Cookies and similar devices

Whilst the DPA will apply to cookies or any similar devices (including 'clear gifs and other spyware') which amount to or contain personal data, the PEC Regulations will also apply to all cookies and similar devices. Regulation 6(2) of the PEC Regulations provides that, subject to a few exceptions,[66] cookies and similar devices cannot be used unless the subscriber or user of the relevant terminal equipments is:

- provided with clear and comprehensive information about the purposes of the storage of, or access to, that information; and

- given the opportunity to refuse the storage of, or access to, that information.

Thus privacy/cookie policy information should provide the above information even where one is certain that no personal data is involved.[67] It is vital that such information be given in a very prominent, clear and comprehensible form so that all users and subscribers (not just the technically literate) are capable of understanding and acting on it. The ICO's Privacy Guidance[68] states that a link to a cookie or privacy policy alone is unlikely to be sufficient unless the relevant section is clearly signposted when you arrive at the page.[69]

5.5 Direct marketing

The PEC Regulations apply to the sending of direct marketing messages by electronic means such as by telephone, fax, email, text message and picture (including video) message and by using an automated calling system.

What is the definition of direct marketing?

Section 11 of the DPA refers to direct marketing as 'the communication (by whatever means) of any advertising or marketing material which is directed to particular individuals'.

As is the spirit of the Act, the definition covers a wide range of activities including the offer for sale of goods or services and also the promotion of an organisation's aims and ideas. All types of organisations including charities or political parties are caught and activities such as appeals and campaigns for support would fall under the definition.

---

66 Such as where storage or access is strictly necessary for the provision of an information society service requested by the user or subscriber.
67 Regulation 6(2) provides that so long as the required information was given the first time such a device was used to store or access data in the terminal equipment of a user or subscriber then that information will not need to be given on subsequent occasions.
68 ICO Guidance on the Privacy and Electronic Communications (EC) Directive Regulations 2003. Part 1.
69 The Privacy Guidance issued by the ICO states that where devices are employed by one party on another's website (eg though a third party advert) then both parties will have responsibilities (further detail on this topic including on the relevant exceptions is available in the ICO's Privacy Guidance (cited above)).

Non-electronic communications

Unless a subscriber has previously consented, an organisation is not allowed to contact the subscriber using an automated calling system for direct marketing purposes (Regulation 19). The position is slightly different when looking at non-automated systems. These will not be permitted in the case where a subscriber has either notified the organisation that they should not be called ('opt-out'), or they have notified OFCOM that he/she wishes to be included in an opt-out register. Individuals can register with the Telephone Preference Services ('TPS') for free which is run by the Direct Marketing Association. The TPS is the central opt-out register and once an individual has recorded a desire not to receive unsolicited calls, organisations have a legal requirement to comply with that wish unless consent is otherwise given.

Organisations which engage in direct marketing should check the TPS register before making unsolicited telemarketing calls to individuals. Conversely, the use of fax machines for direct marketing purposes is not permitted in the case of an individual subscriber unless he or she has previously consented ('opt-in') to receive materials. Changes in 2004 now mean that corporate subscribers can now register with the TPS[70].

5.6 E-mail marketing

The PEC Regulations provide that an organisation cannot transmit, or instigate the transmission of, unsolicited (i.e. not actively invited, either directly or through a third party) marketing material by electronic mail ('e-mails') (note this includes not only what is usually considered to be an e-mail but also texts/SMS messages, voice messages and various other categories) to an individual subscriber, unless that subscriber has previously notified the sender that they consent for the time being. Subject to certain exceptions, individual subscribers must opt in to receive such e-mails. There is, though, an exception to the need for formal 'opt-in' consent (commonly called a 'soft opt-in'), which applies where the direct marketer has:

* obtained contact details of the recipient in the course of a sale or 'negotiations' for a sale of products or services to that recipient[71];

* the direct marketing is conducted in respect of similar products and services[72];

* the recipient has been given a simple means, without charge, at the time of initial collection of the data, to refuse (or 'opt-out') of the use of his contact details for direct marketing; and

* included in each subsequent e-mail to an individual is a right to 'opt-out' of such future direct marketing (PEC Regulations Regulation 22).

Clearly 'soft opt-in' cannot work with purchased lists and the scope is narrow. In relation to lists provided by third parties or where the 'soft opt-in' option otherwise

---

70 Privacy and Electronic Communications (EC Directive/Amendment) Regulations 2004 SI 2004/639.
71 The ICO's Privacy Guidance contains detailed commentary on the scope of 'negotiation'. For example, whilst an active expression of interest or a request for a quote might, in the circumstances, amount to 'negotiation', the use of cookie technology to identify a user's interests would not be sufficient.
72 The ICO's Privacy Guidance suggests that 'similar' should be interpreted with a view to what a customer would reasonably expect to receive further information on.

cannot be used, 'consent'[73] will be necessary (it may, anyway, be safest to always ensure there is explicit opt-in consent). However, 'corporate subscribers' do not benefit from the protections against unsolicited e-mail and nor do these provisions cover their employees using corporate e-mail systems for personal purposes.[74]

The ICO has issued a Good Practice Note in relation to electronic mail marketing.[75]

5.7 Calling Line Identity (CLI) Information

The PEC Regulations also impose obligations upon a communications service provider in relation to CLI information. CLI information is that information displayed on call return and call display facilities such as the information which alerts a subscriber to the identity of an incoming caller and the '1471' service. The Regulations impose privacy obligations in relation to both outgoing and incoming call information.

Outgoing Calls—Regulation 10

The service provider must provide a user with a simple means of preventing the presentation of the identity of his calling line when that call is connected to another line.

Incoming Calls—Regulation 11

The service provider must also ensure that a user has a simple means of ensuring that the identity of a calling line is not shown on connection of the call to him. This is used in cases where the caller's anonymity needs to be guaranteed, such as help lines for alcoholics and children. This obligation extends to call forwarding, such as where an out of hours call is forwarded from a doctors' surgery to that doctor's home phone. The service allows that home number to remain private. Where someone receives a call with no CLI information shown, the service provider is obliged to provide that recipient with a simple means of rejecting the call.

The service provider must provide users with these means of protecting their CLI information free of charge.

---

73 As with the DPA, consent requires some from of active communication (such as a ticked box or an e-mail or subscribing to a service). The individual must likewise fully understand the consequences of their consent. Privacy Guidance issued by the ICO states that a failure to object (ie failure to tick a box stating one objects to use of information for marketing) will not amount to consent (although it may form part of a wider nexus demonstrating consent if sufficiently clear).

74 ie companies, limited liability partnerships, Scottish partnerships, corporations sole and any other body corporate or entity which is a legal person distinct form its members. The ICO Privacy Guidance makes clear that marketers must obtain prior consent of recipients to send unsolicited e-mail to any e-mail addresses used by a non-limited liability partnership. Guidance also states that whilst PEC REGS may not apply in this context to protect employees of corporate subscribers from unwanted mail, the provisions of the DPA may be relevant.

75 http://www.ico.gov.uk/upload/documents/library/data_protection/practical_application/electronic_mail_marketing_12_06.pdf.

Exceptions—Regulations 15 and 16

The service provider does not have to prevent CLI information on outgoing calls to the 999 number, to enable the emergency services to respond effectively to calls by knowing the caller's location. Furthermore, a service provider is not restricted under the Regulations from assisting with tracing malicious or nuisance calls to a user who has requested the tracing of calls to his or her line. The CLI information can be made available to that user.

6 Other Legislation on Data Protection or Data Retention

6.1 The EU Data Retention Regime

The EU regime regulating the retention of communications data stems from the Data Retention Directive 2006/24/EC (the 'Directive').

The Directive came into force on 3 May 2006. Member States had until 15 September 2006 to implement the Directive, save that Member States could postpone application of the Directive to 'the retention of communication data relating to Internet Access, Internet telephony and Internet e-mail' for up to 18 months (i.e. until 15 March 2009) [Art 15.3].

The United Kingdom elected to do just that, recognising that the retention of communications data relating to the Internet was a more complex issue, involving larger volumes of data and a broader set of stakeholders than the retention of data arising from fixed-line and mobile telephony services.

The objective behind or reason for retaining the relevant data is

'... in order to ensure that the data are available for the purpose of the investigation, detection and prosecution of serious crime, as [that term] is defined by each Member State in its national law' [Art 1.1].

The Directive provides that Member States shall adopt measures to ensure that the data is;

'provided only to the competent national authorities in specific cases and in accordance with national law' [Art 4]. The Directive applies to 'providers of publicly available electronic communications services or of a public communications network' [Art 3.1].

Retention of data

The Directive provides that the relevant data may be 'retained for periods of not less than six months and not more than two years from the date of the communication' [Art 6].

Article 7(d) of the Directive provides that retained data 'shall be destroyed at the end of the period of retention', but, somewhat curiously, includes a carve-out for 'those [data] that have been accessed and preserved'. [Art 7(d)] The Directive does not elaborate on exactly what is meant by this. For example, is the intention that it is the responsibility of the service provider to retain in perpetuity any data that has been 'accessed'? It would seem to make more sense that it should be the relevant competent national authority that has requested access to the data who is responsible for the ongoing safe-keeping of such data.

There are other issues with the language used in the Directive. For example, it is noted that Article 5.1(c)(2)(i) refers to 'the IP address ... allocated by the Internet access service provider to a communication', whereas, generally speaking, IP addresses are allocated to particular machines or end users, not individual communications or emails.

Importantly, it is clear that the Directive applies to 'the content of electronic communications, [which includes] information consulted using an electronic communications network' [Art 1.2]. That is, no data revealing the content of a communication may be retained. This means, importantly for ISPs, that no data on web pages visited is required to be retained.

However, the Directive is not prescriptive with regard to the gaining of access to, and use of, the retained data by public authorities and/or law enforcement authorities of the Member States. This is left for Member States to deal with under their national laws.

Security of data

The Directive sets out a minimum set of standards that must be adhered to with respect to the security of the retained data, namely as follows:

'(a) *the retained data shall be of the same quality and subject to the same security and protection as those data on the network;*

(b) *the data shall be subject to appropriate technical and organisational measures to protect the data against accidental or unlawful destruction, accidental loss or alteration, or unauthorised or unlawful storage, processing, access or disclosure;*

(c) *the data shall be subject to appropriate technical and organisational measures to ensure that they can be accessed by specially authorised personnel only; and*

(d) *the data, except those that have been accessed and preserved, shall be destroyed at the end of the period of retention.'* [Art 7]

The Directive does not require Member States to reimburse service providers for the cost of compliance. Instead, Member States are free to decide whether or not they will compensate service providers. One of the key drivers behind the Directive, as stated in the first Article of the Directive, was supposedly

'to harmonise Member States' provisions concerning the obligations of ... providers ... with respect to the retention of certain data which are generated or processed by them' [Art 1.1].

Arguably, however, the Directive fails to achieve this objective. Consider, for example, the economic effect on different service providers located in different Member States of:

- variations in the length of the retention period (6 months to 2 years);

- variations relating to the reimbursement, or otherwise, of service providers for additional costs incurred; and

- variations in each Member State's interpretation of the somewhat vague terms used in relation to data generated in connection with Internet services.

6.2 UK transposition of the Directive

On 1 October 2007 the Data Retention (EC Directive) Regulations 2007 came into force, forming the initial transposition of the Directive into UK law, and only applied to providers of fixed-network and mobile telephony services. The complete transposition of this Directive into UK law has taken three years to achieve and was completed by way of the Data Retention (EC Directive) Regulations 2009 (the '2009 Regulations'). The 2009 Regulations replaced and extended the scope of the 2007 legislation so that they now encompass internet communications data.

The Secretary of State must give written notice to the public communications provider if the Regulations are to apply to it, unless the service provider's data is already being retained in the United Kingdom in accordance with the Regulations by another public communications provider [Reg 10(2)]. That is, there should be no duplication of storage (and associated costs) of retained data. If data is retained upstream by suppliers of services, there should be no need for resellers to retain data, provided they have suitable arrangements regarding data retention in place with their wholesalers.

Application and scope

The 2009 Regulations are expressed to apply to all 'public communications providers', a phrase which is defined to mean a provider of a 'public electronic communications network' or a 'public electronic communications service', as those terms are defined in section 151 of the Communications Act 2003 [Reg 2 (e)].

The 2009 Regulations provide for a retention period of 12 months in the UK [Reg 5]. Unfortunately, the 2009 Regulations contain the additional language that appears in Article 7(d) of the Directive, which presumably means that in the UK, there remains some uncertainty as to whether all data can be destroyed at the end of the 12 month retention period [Reg 6(1)(d)].

The data that is required to be retained is set out in the Schedule to the Regulations and mirrors the content and language contained in the Directive. The Schedule is divided into three parts for each of 'fixed network telephony', 'mobile telephony' and 'internet access, internet email or internet telephony'.

The categories of information to be retained are reasonably comprehensive and, as they say, the devil is in the detail, particularly with regard to the precise meanings of some of the words and phrases used. What will be of interest to ISPs and the like is the interpretation of what might at first seem like relatively mundane terms contained in the Regulations, such as 'email' (for example, does this include instant messenger services?). However, as the Regulations don't deal with this point, we will have to wait until the next instalment of guidance issued by the Information Commissioner to see how such terms are to be interpreted.

The language issue relating to Article 5.1(c)(2)(i) has not yet been resolved and remains in the Regulations at paragraph 13 (b) of the Schedule. Importantly, it is clear from the Regulations (see Reg 4(5)) that as with the Directive, no data revealing the content of a communication may be retained.

The Regulations provide that the data must be able to be transmitted by the service provider to the competent authorities 'without undue delay' [Reg 8].

Thus, not only will service providers be required to store large amounts of data, they will also need to ensure that they can promptly identify and retrieve the data following a request from a competent national authority.

There is no indication of what might be an acceptable period of delay (hours, days, weeks or months) or if such period is flexible in response to, for example, when a request is received or the number of requests received. Consider, for example, a request made at 9am on Good Friday or whether there is a requirement for service providers' data storage and retrieval systems to be infinitely scalable.

Access and security

Again, the Regulations do not deal with the issue of access to retained data, because access to the data is dealt with under other relevant legislation, including the Regulation of Investigatory Powers Act 2000 and the Anti-Terrorism, Crime and Security Act 2001. Obviously, however, the issue of who has access to the retained data is very relevant as there is likely to be a direct correlation between the number of competent national authorities that may request access to the data and the cost to service providers of providing data in response to such requests, particularly given the requirement that the data retained can be transmitted without undue delay in response to requests. There is also the outstanding issue of who else may obtain access to the retained data and under what circumstances. For example, it is not clear whether a court is able to grant access in connection with a civil matter (this is not currently expressly prohibited by the 2009 Regulations).

The minimum standards set out in Article 7 of the Directive are repeated in Regulation 6 of the 2009 Regulations, essentially word-for-word, including the uncertain language in Article 7(d) regarding 'those [data] that have been accessed and preserved', which is repeated as 'lawfully accessed and preserved' (Reg 6(1) (d)).

The data-security provisions in Article 7/Regulation 6 fall short of the recommendations made by the Article 29 Data Protection Working Party. The Working Party recommended, for example, that systems for storage of data for public-order purposes should logically be separated from systems used for business purposes. The Working Party consistently resisted the introduction of the Directive on grounds of privacy and the protection of individuals' fundamental rights. Given that the Working Party's recommendations are not binding, it is not surprising that its suggestions did not find much favour with those Member States that have supported wide-ranging data retention provisions from the outset, such as the United Kingdom.

Reimbursement of costs

Regulation 11 provides for reimbursement of the cost of compliance to service providers, particularly that:

- '(1)   The Secretary of State may reimburse any expenses incurred by a public communications provider in complying with these Regulations.'

- (2)   Reimbursement may be conditional on the expenses having been notified to the Secretary of State and agreed in advance.'

- (3)   The Secretary of State may require a public communications provider to comply with any audit that may reasonably be required to monitor reimbursement.

Obviously, reimbursement of a service provider's increased costs is an additional incentive to encourage compliance. It remains to be seen, however, how generous the Government will be in the future, given that compliance with the draft Regulations will be mandatory, whereas the Government's obligation to reimburse is voluntary.

Every year, the communications provider must also provide the Secretary of State with statistical information including the number of times retained data were disclosed in response to a request, the time lapse between requests and disclosure and the instances when disclosure could not be met (Reg 9).

6.3   Recent Developments

In response to concerns as to the scope of the new Regulations, the government has set up an 'implementation group' to support the Home Office's work in providing guidance on the new regime. The status and membership of this group remains undecided, although any assistance it can provide in interpreting the 2009 Regulations will no doubt be warmly welcomed by the communications sector. Because the Regulations are so new, it remains to be seen what approach the Information Commissioner and the courts will take to their interpretation and application.

7   Breach of the DPA

7.1   Breach of the principles

Compliance with the DPA is principally monitored by the ICO. If the Commissioner becomes aware of a breach of the DPA (for example as a result of a complaint from a member of the public) he will carry out an assessment and where appropriate take direct action against the person responsible. The ICO may take the following actions against a data controller:

Fines

At present the ICO does not have the power to impose a fine or other penalty against a data controller following breach of the security requirements under the DPA, however serious. Instead, the ICO is likely to issue (or threaten to issue) a formal enforcement notice against the data controller.

Enforcement notices

An enforcement notice is a formal notice issued by the ICO, requiring the relevant data controller to take a specific course of action to ensure compliance with the DPA.

Data controllers that fail to comply with an enforcement notice may be exposed to criminal prosecution (including, in certain circumstances, the directors and officers of the organisation concerned[76]). The offence carries a maximum penalty of a £5000 fine in the Magistrates Court and an unlimited fine in the Crown Court.

Voluntary undertakings

The ICO can use its discretion to accept undertakings in lieu of an enforcement notice from the data controller. These are voluntary admissions that a lapse has taken place, supported by undertakings to remediate the failings and prevent a recurrence. Failure to comply with an undertaking will lead to an enforcement notice being issued.

Note that the ICO will make undertakings publicly available and will not accept undertakings on the condition that they be kept confidential.

7.2 Security breaches that involve the data processor

Unless a security lapse has arisen due to the data processor committing an offence under section 55 of the DPA (offences of unlawfully obtaining or disclosing personal data without the consent of the data controller), the ICO won't have power to take formal action against a data processor. Rather it will pursue the relevant data controller who will be expected to demonstrate that it had carried out proper due diligence when selecting the data processor regarding the security arrangements to be deployed in processing the personal data and appropriate taken steps subsequently to monitor and manage compliance with those standards of care.

Assuming a contract has been put in place between the data controller and data processor consistent with the seventh principle requirements, the lapse is likely to give rise to a breach of contract claim by the data controller against the data processor.

---

76   Section 61, DPA.

7.3 Scope of the ICO's powers

The ICO can only impose a fine if the terms of an enforcement notice are breached—to date few of these have been issued. The main weapon is damage to reputation. The ICO is not shy of publishing enforcement notices and voluntary undertakings to 'name and shame' culprits. In the current climate, alleged security failings generate intensive media scrutiny and will have an adverse impact on the data controller.

Whilst the ICO's powers are currently limited, with passage of the Criminal Justice and Immigration Act 2008 ('CJIA') on 8 May 2008, there is the prospect of much greater enforcement powers coming into effect very shortly.

Specifically, the CJIA grants the ICO powers to impose Monetary Penalty Notices on data controllers who breach a provision in the Act if it is satisfied that:

- there has been a serious contravention of one of the DPA principles, including the seventh principle;

- the contravention was of a kind likely to cause substantial damage or substantial distress; and

- the contravention was either deliberate or the data controller knew, or ought to have known, that there was a risk that the contravention would occur and it would be likely to cause substantial damage or substantial distress, but failed to take reasonable steps to prevent it.[77]

It remains to be seen how widely these powers will be used or the level of fines that will be levied.

7.4 Enforcement trends

In the past the ICO has usually approached data controllers with a view to ensuring compliance before commencing enforcement action. This approach was emphasised in the enforcement strategy document published by the ICO in November 2005 and was part of an approach which centred on encouraging compliance rather than issuing punishment. However, more recently the ICO's tendency to issue preliminary enforcement notices at an early stage in its investigations reveals that data controllers can no longer be sure of the ICO continuing to take a reconciliatory approach.

7.5 Relationship between the ICO and the FSA

If a data controller is also regulated by the FSA, then the regulators will co-ordinate investigations into breaches of data security between themselves. In the past the ICO has taken a lead in handling breaches, but then passed findings to the FSA to take enforcement action. This reflects the fact that the FSA has to date had far more extensive enforcement powers over regulated financial bodies. It remains to be seen whether this approach will continue given new powers to be granted to the ICO under the CJIA. It may be that we will see the ICO taking more of a leading role in the future.

---

77   Section 144, CJIA.

8   Recent developments

8.1   BSI Standards

In June 2009, the BSI published a new standard[78] relating to compliance with the DPA. The objective of this British Standard is to enable organisations to put in place a personal information management system ('PIMS'), thus improving compliance with data protection laws. Its focus is primarily placed on ensuring that organisations provide sufficient guidance and resources rather than prescribing a specific operation system. It is hoped that small to medium sized businesses will benefit from the standard which will be more accessible than the detailed provisions of the DPA.

8.2   Bill to amend DPA published

On 14 January 2009 the Ministry of Justice published the Coroner's and Justice Bill. The bill proposes new provisions to increase the ICO's power to perform spot checks on public authorities and impose deadlines on data controllers for the submission of information necessary for the ICO to assess their compliance with the DPA. The bill also makes changes to the data sharing conditions in the DPA and provides a procedure for removing or modifying barriers to data sharing. The bill proposes to place a new obligation on the ICO to publish a data sharing code of practice and changes to the way that the ICO is funded are also proposed.

Increased powers for the ICO

The bill seeks to introduce a new right for the Information Commissioner to serve a government department or a designated public authority with an assessment notice so that compliance with the DPA principles can be established. A data controller, on receiving such a notice, will be required to permit the commissioner to enter any specified premises for the inspection or examination of any documents, information, equipment or materials that the commissioner wishes to view.

Further powers granted to the ICO

The powers relating to information notices served under the DPA have been strengthened so that the data controller will have to, in the future, provide information within a specified period at a specified time and place and in a specified form. The bill also proposes to extend the powers for warrant for entry and inspection. In particular it gives the Information Commissioner the power to require any person on the premises to provide an explanation of any document or other material found on the premises.

Data sharing

The bill proposes that certain designated authorities should now be able to make an order requiring information sharing. The designated authorities specified under the

---

78   BSI Standard BS 10012:2009 available at www.bsigroup.com.

bill are stated to be the Scottish ministers, the Welsh ministers, a Northern Ireland department, the Secretary of State, the Treasury or any other minister in charge of a government department. Under the new proposals the order must set out the personal class of persons permitted to share the information and the purposes for which the information is shared. If sharing is found to be necessary to secure a relevant policy objective; is proportionate to that policy objective; and strikes a fair balance between the public interest and any individual affected, the designated authority may share that information.

The new data sharing proposals have attracted significant controversy as they increase the level of information sharing between government departments which had not been previously envisaged by the DPA.

8.3   EC Working party proposes amendments to E-Privacy Directive

In February 2009, the EC Article 29 Data Protection Working Party published an opinion on the proposals to amend E-Privacy Directive 2002/58/EC ('E-Privacy Directive'). The opinion has been passed in light of reform proposals for EU Telecoms regulations (discussed further in chapter 1). The Working Party had made a number of previous recommendations which had been taken into account during the legislative process. However, it remained a concern that a number of issues had been disregarded. In particular, the Working Party raised concerns over changes to be made to the way that security breaches were dealt with under the E-Privacy Directive and also raised opposition to the parliament's amendment that default browser settings should mean that an individual has provided prior consent. On 6 May 2009, the European Parliament adopted a legislative resolution in its second reading of the European Commission's proposal to amend the E-Privacy Directive. That resolution takes account of the Working Party's comments as well as the European Council's common position on the Commission's proposal. The Parliament's resolution will now be sent to the Council for a second reading. Although their proposed amendments to the E- Privacy Directive are unlikely to be challenged, other amendments to telecoms legislation under the resolution may not be accepted by the Council and as a result the whole package of amendments appears likely to be referred to the Conciliation Committee. Further details of the Telecoms Framework Review are contained in Chapter 1.

Chapter 3

General Conditions of Entitlement

Michael H. Ryan and Simon Cloke
Arnold & Porter (UK) LLP*

ARNOLD & PORTER (UK) LLP

3.1 Authority to Provide Networks, Services and Associated Facilities

Under the provisions of the EU regulatory framework for electronic communications, undertakings may be required to obtain a general authorisation before providing electronic communications networks or services.[1] No such authorisation requirement has been adopted in the United Kingdom. Prior to 25 July 2003, a person running a telecommunication system within the United Kingdom was required to obtain a licence from the Secretary of State for Trade and Industry.[2] This requirement was repealed upon the entry into force of the Communications Act 2003[3] and was in effect replaced by a requirement that a person intending to provide an electronic communications network, service or associated facility in the United Kingdom that has been 'designated' by OFCOM for that purpose must notify OFCOM in advance of that intention.[4] OFCOM has made no such designations to date, with the result that there is at present no requirement to notify or obtain any form of authorisation before a person may provide electronic communications networks, services or associated facilities in the United Kingdom.[5]

While there is no authorisation requirement in the United Kingdom, providers of 'electronic communications networks,' 'electronic communications services' and 'associated facilities' must comply with certain conditions set by OFCOM. These conditions include the general conditions (referred to by OFCOM as 'General

---

*   Solicitors, Tower 42, 25 Old Broad Street, London, EC2N 1HQ, England. Tel: +44 (0) 20 7786 6100.
     e-mail: michael_ryan@aporter.com.
1   Authorisation Directive, 2002/20/EC, OJ L 108, 24.4.2002, p. 21, article 3, 2.
2   Telecommunications Act 1984, section 7 [repealed].
3   See Communications Act 2003, section 147. This provision implemented the Authorisation Directive,
     article 3, which has the effect of prohibiting Member States from imposing an obligation on operators
     to obtain an individual licence as a precondition to providing service except in specified cases.
4   Communications Act 2003, section 33. Article 3,2 of the Authorisation Directive provides that the
     provision of services may be made subject to a general authorisation only and that operators may be
     required to notify their intention to provide services.
5   Note, however, that the use of a radio station or radio apparatus for the provision of electronic
     communications services may require a wireless telegraphy licence—see Chapter 7.

Conditions of Entitlement') described in this chapter. Some providers must, in addition, comply with other conditions described in Chapter 4.

An 'electronic communications network' is defined in the Communications Act 2003 as 'a transmission system for the conveyance, by the use of electrical, magnetic or electro-magnetic energy, of signals of any description' and includes the apparatus comprised in the system; apparatus used for the switching or routing of signals, and software and stored data.[6] 'Electronic communications service' is defined in corresponding terms[7] except that 'content services'[8] are expressly excluded. Taken together, these definitions embrace all types of service platform (fixed, mobile, satellite, internet, ...) and all types of service provided over those platforms (voice, data, text, SMS, fax, ...) except (as required by the exception for 'content services') to the extent that a service consists in 'the provision of material with a view to its being comprised in signals conveyed by means of an electronic communications network,' and/or 'the exercise of editorial control over the content of signals conveyed by means of such a network.' Although the precise boundaries of what constitutes a 'content service' may not always be clear, the exception for such services is generally understood to remove broadcasting and similar internet-based services from the scope of the General Conditions of Entitlement discussed in this chapter and the other conditions discussed in Chapter 4.

'Associated facility' includes facilities that are available for use in association with the use of electronic communications networks or services (whether or not provided by the person making the facility available); that makes the provision of that network or service possible; or that makes possible the provision of other services provided by means of that network or service or supports the provision of that service.[9] Examples of 'associated facilities' include conditional access systems[10] and electronic programme guides.[11]

3.2 General Conditions

OFCOM has the power to set general conditions[12] relating to a broad range of matters defined in the Act.[13] There are currently twenty-two General Conditions of

---

6 Communications Act 2003, section 32(1).
7 Communications Act 2003, section 32(2).
8 Communications Act 2003, section 32(7).
9 Communications Act 2003, section 32(3).
10 Concerning conditional access systems, see § 4.3.
11 Concerning electronic programme guides, see § 4.2.
12 Communications Act 2003, section 45(2)(a).
13 The matters that may be made the subject of general conditions are defined in Communications Act 2003, sections 45(3), 51, 52, 57, 58 and 64. These provisions implement the Authorisation Directive, article 6, which provides that general authorisations (see footnote 7 and the accompanying text) may be made subject only to the conditions listed in Annex I of the Directive. Such conditions must be objectively justified in relation to the network or service concerned, non-discriminatory, proportionate and transparent.

Except where explicitly stated, the basis for all of the General Conditions is article 6 of the Authorisation Directive.

Entitlement.[14] OFCOM may make these conditions applicable (i) to every person providing electronic communications networks or services, or (ii) to every person providing a network or service of a particular description.[15] The general conditions set by OFCOM typically apply to 'Communications Providers', but that term is defined differently in different conditions. In order to determine whether a particular condition applies to a particular network or service provider, the definition used in each condition must be examined. Broadly speaking, there are three categories of Communications Provider distinguished for these purposes:[16]

- Providers of electronic communications services ('ECS') or networks ('ECN')

- Providers of public electronic communications services ('PECS') or networks ('PECN')

- Providers of publicly available telephone services ('PATS') or public telephone networks ('PTN').

The relationship between these three categories of Communications Provider is represented in Figure 3.1 and explained below.

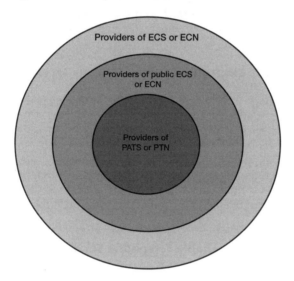

Figure 3.1—Types of Provider

14 OFTEL, *Notification setting General Conditions under section 45 of the Communications Act 2003*, 22 July 2003, introduced 21 of these General Conditions of Entitlement (some of which have been subsequently amended). The twenty-second was set in OFCOM, *Broadband migrations: enabling consumer choice*, 13 December 2006.
    OFCOM has proposed amendments to several General Conditions. Concerning General Conditions 8, 14 and 19, see OFCOM, *Telephone directory information obligations and regulations*, 10 March 2008; concerning General Condition 17, see OFCOM, *Changes to 0870*, 2 May 2008; concerning General Condition 14, see OFCOM, *Review of Alternative Dispute Resolution and Complaints Handling Procedures*, 10 July 2008.
    OFCOM has proposed the introduction of a new General Condition 23 imposing requirements upon all communications providers providing a mobile service in respect of sales and marketing of such services to domestic and small business customers: see OFCOM, *Protecting consumers from mis-selling of mobile telecommunications services*, 18 March 2008. General condition 23 enters into force on 16 September 2009.
15 Communications Act, section 46(2).
16 See OFTEL, *Notification setting General Conditions under section 45 of the Communications Act 2003*, 22 July 2003, Schedule 1. Some of these conditions have since been amended. Where this is the case, this is indicated in the discussion of each condition which follows.

Providers of electronic communications services or networks

The first category of Communications Provider (represented by the outer ring in Figure 3.1) is the widest and includes *all* persons that provide electronic communications networks or services, including public and private networks, fixed and mobile voice telephony (including VoIP), data and internet service providers, and operators that own their networks as well as resellers (including all persons in the two inner rings). Although this category is very broad, it does not include terminal equipment providers.

The General Conditions of Entitlement draw no distinctions based on ownership of the network. Providers of electronic communications networks and providers of electronic communications services that lease all or parts of their networks from another operator may both qualify as Communications Providers. Thus, even system-less resellers of ECS are providers of ECS. The 'provider' of an electronic communications network for the purposes of a condition is the organisation which has control over the facilities to which the condition relates. The provider is typically the organisation that has the contract with the end-user, or where the service is provided on a wholesale basis, with the reseller or intermediary.[17]

Providers of Public Electronic Communications Services or Networks

The second category is a subset of the first. It includes only providers of 'public' electronic communications services or networks. A 'public electronic communications service' is an electronic communications service that is provided for use by members of the public. A network provided wholly or mainly for the purpose of providing public electronic communications service is referred to as a 'public electronic communications network.' Organisations that provide services or own networks that are not made available to the public (such as private networks or bespoke services not offered to the public) are excluded from this category.

Providers of Publicly Available Telephone Services or Public Telephone Networks

The third category is a subset of the second and includes only persons that provide a 'publicly available telephone service' or 'PATS'; i.e. a service available to the public for originating and receiving national and international calls and access to emergency organisations through a number or numbers in a national or international telephone numbering plan.[18] This category includes conventional fixed, mobile voice telephone networks, indirect access services and resellers of

---

17 OFCOM, *General conditions of entitlement (guidelines)*, undated. http://www.OFCOM.org.uk/telecoms/ioi/g_a_regime/gce/gcoe/—pages 1 and 2.

18 OFCOM, *Regulation of VoIP services*, 29 March 2007, paragraph 3.12. The definition in the Notification of 22 July 2003, *supra*, adds that PATS:

'may, where relevant, include one or more of the following services: the provision of operator assistance services, Directory Enquiry Facilities, Directories, provision of Public Pay Telephones, provision of service under special terms, provision of specific facilities for End-Users with disabilities or with special social needs and/or the provision of non-geographic services.'

For a discussion of how OFCOM defines PATS, see *Regulation of VoIP services*, 29 March 2007, Annex 5, paragraphs 5.17 *et seq.*

PATS. It also includes voice over internet protocol (VoIP)[19] if the service satisfies the preceding definition of PATS.[20] This category excludes data services such as data networks and internet access services.

PATS is specially defined for the purposes of General Condition 18—see below.

Figure 3.2 summarises the general conditions that are applicable to each of these three categories of service and network provider. A short description of each of the twenty-two general conditions follows in §§ 3.2.1 *et seq.*

Practical Issue: VoIP

VoIP voice call services may, depending on their individual characteristics, fall into different categories. OFCOM has distinguished four types of VoIP voice call service for regulatory purposes:

Type 1: peer-to-peer services to make and receive voice calls over the Internet only, usually within the same application community;

Type 2: VoIP Out services to make voice calls over the Internet to the PSTN (public switched telephony network), but not to receive calls from the PSTN;

Type 3: VoIP In services to receive voice calls over the Internet from the PSTN, but not to make calls to the PSTN. Customers can be allocated an ordinary geographic number or a VoIP number (056); and

Type 4: VoIP In and Out services to receive voice calls over the Internet from the PSTN and to make voice calls over the Internet to the PSTN. Customers can be allocated an ordinary geographic number or a VoIP number (056).

Depending on their individual characteristics, OFCOM considers Type 1 VoIP services are unlikely to constitute an ECS. Type 2 and Type 3 VoIP services are likely to be regarded as PECS. Type 4 VoIP services are likely to be PECS or, if they meet the PATS gating criteria, PATS.[21]

3.2.1   General Condition 1—General Access and Interconnection Obligations

Paragraph 1.1 of General Condition 1 requires persons who provide a Public Electronic Communications Network to negotiate interconnection agreements with other EU communications providers within a reasonable time on a good faith basis.[22]

---

19 Following OFCOM's usage of the term in *Regulation of VoIP services*, 29 March 2007, footnote 40, 'VoIP' is used here in a broad sense to include Voice over the public Internet, Voice over broadband (including managed and unmanaged services), Voice over Unlicensed Wireless Access, Voice over licensed wireless including pre WiMax based services. It also includes voice services described as 'new voice services' (or 'NVS') in OFCOM's consultation document entitled *New Voice Services: A consultation and interim guidance*, 6 September 2004.

20 See OFCOM, *Regulation of VoIP services*, 29 March 2007, paragraph 3.13. A VoIP service may provide either call origination or call termination or both outside the PSTN, in which case the service is not PATS.

21 OFCOM, *Regulation of VOIP Services: Access to the Emergency Services*, 5 December 2007, paragraphs 3.5 and 3.19.

22 See OFTEL, *The General Conditions of Entitlement, Final statement issued by the Director General of Telecommunications*, 9 July 2003, paragraph 3.5.

General Condition	Providers of Publicly Available Telephone Services or Public Telephone Networks	Providers of Public Electronic Communications Services or Networks	Providers of Electronic Communications Services or Networks
GC. 1 – General access and Interconnection obligations	✓	✓	Part 1.2 and 1.3 only.
GC. 2 – Standardisation and specified interfaces	✓	✓	✓
GC. 3 – Proper and effective functioning of the network	✓	✓	
GC. 4 – Emergency call numbers	✓		
GC. 5 – Emergency planning	✓		
GC. 6 – Public pay telephones	Applies to providers of public pay telephones		
GC. 7 – Must carry obligations	Applies to providers of "Appropriate networks" used for receiving TV.		
GC. 8 – Operator assistance, directories and directory enquiries	✓		
GC. 9 – Requirement to offer contracts with minimum terms	✓	✓	
GC. 10 – Transparency and publication of information	✓		
GC. 11 – Metering and billing	✓	Para. 11.1 and 11.2 only	
GC. 12 – Itemised bills	✓		
GC. 13 – Non-payment of bills	✓		
GC. 14 – Codes of practice and dispute resolution	✓	✓	
GC. 15 – Special measures for end users and disabilities	✓		
GC. 16 – Provision of additional facilities	✓		
GC. 17 – Allocation, adoption and use of telephone numbers	✓	✓	✓
GC. 18 – Number portability	✓	✓	✓
GC. 19 – Provision of directory information	✓	✓	✓
GC. 20 – Non geographic numbers	✓	✓	✓
GC. 21 – Quality of service	✓		
GC. 22 – Service Migrations	✓		
Applies to providers of "Broadband Services"			

Figure 3.2—General Conditions applicable by Type of Service

58

In *Hay Systems Limited (HSL) and T-Mobile (UK) (T-Mobile)*,[23] OFCOM ruled that the obligation in paragraph 1.1 to negotiate interconnection does not impose an obligation to negotiate a particular technical means of providing interconnection. HSL had submitted a written request to T-Mobile for direct SS7 interconnect via Signalling System 7 (SS7) for the purpose of SMS termination. In response to this request T-Mobile suggested that an IP-based wholesale bulk SMS product might be more suitable. In the discussions that followed HSL did not provide T-Mobile with its exact requirements for SS7 nor explain why the IP based solution suggested by T-Mobile was not suitable for its requirements. Having failed to reach an agreement for SS7 interconnection, HSL requested that OFCOM resolve the dispute. OFCOM concluded that the offer of interconnection via IP satisfied the obligation to negotiate in good faith with a view to concluding an agreement for a viable form of interconnection.

Paragraphs 1.2 and 1.3 apply to providers of Electronic Communications Networks and Electronic Communications Services. Under these paragraphs any information obtained from another Communications Provider in the course of negotiations for network access or interconnection must be kept confidential and used solely for the purpose for which it was supplied.[24] The requirement to keep such information confidential does not, however, apply when furnishing information to OFCOM.[25]

In *British Telecommunications plc v. OFCOM*,[26] the Competition Appeals Tribunal ('CAT') ruled that BT had contravened General Condition 1.2 by using customer-specific information acquired from another Communications Provider. The information had been provided with a request for the provision of CPS by BT. BT was found to have used the information to make a 'save call' marketing BT's own services to the customer. BT's use of the information was held to be outside the purpose for which the information was transmitted and therefore contrary to General Condition 1.2.[27]

3.2.2 General Condition 2—Standardisation and Specified Interfaces

General Condition 2 requires that all providers of Electronic Communications Networks or providers of Electronic Communications Services must comply with any relevant compulsory standards/specifications listed in Article 17 of the Framework Directive. In the absence of any compulsory standard, they must take account of any voluntary standards/specifications adopted by the European

---

23 OFCOM *Determination to resolve a dispute between Hay Systems Ltd and T-Mobile (UK) about SS7 – based network access for SMS terminations*, 6 November 2006.
24 General Condition 1.2. The language of General Condition 1.2 is derived from Article 4 (3) of the Access Directive.
25 General Condition 1.3.
26 Case no. 1025/3/3/04, 9 December 2004.
27 Prior to the judgment in this case OFCOM had issued BT a Notification of contravention of General Condition 1.2 to BT in relation to its use of information for the purposes of save activities directed at customers who had moved to Wholesale Line Rental (WLR). This Notification was appealed by BT to the CAT (Case No.1040/3/3/04). Following the judgment of the CAT in case 1025/3/3/04, BT withdrew its appeal and OFCOM issued a new Notification of contravention: see OFCOM, *Use of information for marketing activity directed at Wholesale Line Rental ('WLR')* (CW/00739/12/03). See also OFCOM, *Own-initiative about enforcement of General Condition 1.2 regarding losing CPS providers* (CW/00824/04/05); OFCOM, *Complaint from The Carphone Warehouse Croup plc about BT's compliance with General Condition 1.2 in relation to verification calls* (CW/00909/07/06); and OFCOM, *BT complaint against Unicom regarding conduct relating to the transfer of customers between communications providers* (CW/00915/08/06).

Standards Organisation;[28] and, in the absence of compulsory and voluntary standards/specifications, recommendations adopted by the International Telecommunication Union, International Organisation for Standardisation or International Electrotechnical Committee.

3.2.3   *General Condition 3—Proper and Effective Functioning of the Network*

Under General Condition 3, providers of fixed[29] Public Telephone Networks and providers of fixed Publicly Available Telephone Services must take all reasonable steps to ensure:[30]

- the proper and effective functioning of the public fixed telephone network provided by them at all times;

- the availability of the public fixed telephone network and publicly available fixed telephone services provided by them in the event of catastrophic network breakdown or *force majeure*; and

- uninterrupted access to emergency services as part of any publicly available fixed telephone services offered.[31]

3.2.4   *General Condition 4—Emergency Call Numbers*[32]

Providers of Public Electronic Communications Service (excluding those providing a click-to-call service—a 'Type 2' VoIP call—see above for an explanation) must ensure that any end-user can access emergency organisations by using the emergency call numbers '112' and '999' at no charge and, if access is provided through a pay telephone, without having to use coins or cards.

In addition, providers of Public Telephone Networks must, to the extent technically feasible, make caller location information for all calls to the emergency call numbers available to the emergency organisations handling those calls.[33]

3.2.5   *General Condition 5—Emergency Planning*

Providers of Publicly Available Telephone Services and/or Public Telephone Networks must, upon request by relevant government or emergency organisations,

---

28 Directive 2002/21/EC.
29 It has been proposed in the Review of the Regulatory Framework (see Figure 4.3) that this obligation be extended to *all* publicly available telephone services provided over public communications networks and, if adopted, this provision would become applicable to the provision of mobile services.
30 Concerning the application of this condition to VoIP services, see *Guidelines on the application of PATS obligations to VoIP service providers* in OFCOM, *Regulation of VoIP services*, 29 March 2007, Annex 5, paragraphs A5.57 et seq.
31 The language of General Condition 3 is derived from Article 23 of the Universal Service Directive. The fact that the condition applies only to fixed and not mobile networks seems increasingly anachronistic as fixed networks are replaced by mobile ones. This language is expected to change with implementation of the changes proposed after the Review of the Regulatory Framework (see Figure 4.3). See Chapter 1.
32 General Condition 4 was amended in OFCOM, *Regulation of VoIP Services: Access to the Emergency Services*, 5 December 2007.
33 Concerning the application of this condition to VoIP services, see *Guidelines on the application of PATS obligations to VoIP service providers, supra*, Annex A, paragraphs A5.75 et seq.

make and implement arrangements for the provision or rapid restoration of such communications services as are practicable and may reasonably be required in disasters. The Communications Provider is permitted to recover the costs incurred in making or implementing such arrangements (whether from the relevant government or emergency organisation or its customers) and to make the implementation of such arrangements conditional upon being indemnified by the person for whom the arrangements are being implemented.

3.2.6   General Condition 6—Public Pay Telephones

Providers of Public Pay Telephones must comply with obligations relating to access to operator assistance, directory enquiries, display of pricing and other information and wheelchair accessibility.

3.2.7   General Condition 7—Must Carry Obligations

Providers of Appropriate Networks (a term which refers to networks which are used by a significant number of people as their principle means of receiving television programmes[34]) may be required by OFCOM to transmit or broadcast the services of public service broadcasters and to comply with any order made by the Secretary of State concerning the terms on which such material is to be broadcast or transmitted.

The following television services have been included on the 'must carry' list of providers of Appropriate Networks:

- any service of television programmes provided by the BBC so far as it is provided in digital form and is a service in relation to which OFCOM have functions;
- the Channel 3 services so far as provided in digital form;
- Channel 4 so far as provided in digital form;
- Channel 5 so far as provided in digital form;
- S4C Digital; or
- the digital public teletext service.

3.2.8   General Condition 8—Operator Assistance, Directories and Directory Enquiries

Providers of Publicly Available Telephone Services must ensure that any end-user can access operator assistance services and, for a reasonable charge, a directory enquiry facility containing directory information on all subscribers in the United Kingdom who have been assigned Telephone Numbers by any communications provider—except those Subscribers who have exercised their right to have their Directory Information removed.[35]

Providers of Publicly Available Telephone Services who assign telephone numbers to their subscribers must ensure that each of those subscribers is, on request and

---

34   See Communications Act 2003, section 272(7).
35   General Condition 8.1.

for a reasonable charge, supplied with a directory containing directory information on all Subscribers who have been assigned telephone numbers in the Subscriber's local area.[36] Subscribers who require directories containing directory information for subscribers assigned telephone numbers by any communications provider outside of the local area must be supplied such directories by the communications provider on request and for a reasonable charge.[37]

3.2.9   *General Condition 9—Requirement to Offer Contracts with Minimum Terms*

Providers of Public Electronic Communications Services, excluding any broadcast service for television for general and free reception, must, in offering to provide, or providing such services to a consumer, and on the request of that consumer, offer to enter into a contract or vary an existing contract with that consumer which specifies the following minimum requirements:[38]

- the identity and address of the provider;
- the services provided, details of the service quality levels offered and the time for initial connection;
- details of maintenance services offered;
- particulars of prices and tariffs, and the means by which up-to-date information on all applicable tariffs and maintenance charges may be obtained;
- the duration of the contract, the conditions for renewal and termination of services and of the contract;
- any applicable compensation and/or refund arrangements which will apply if contracted quality service levels are not met; and
- the method of initiating procedures for settlement of disputes in respect of the contract.

Where the provider intends to modify a condition in a contract with a consumer which is likely to be of material detriment to the consumer, the provider should:

- provide the consumer with at least one month's notice of its intention detailing the proposed modification; and
- inform the consumer of the ability to terminate the contract without penalty if the proposed modification is not acceptable to the consumer.

3.2.10   *General Condition 10—Transparency and Publication of Information*

A person who provides end-users with access to and use of Publicly Available Telephone Services, except public pay telephones, must ensure that clear and up-to-date information on their applicable prices and tariffs (excluding bespoke or individual prices and tariffs) and on their standard terms and conditions is published, including at least the following:

---

36   The reference to the inclusion of information on subscribers in the 'local area' implies that the directories referenced in this condition need not include mobile numbers.
37   General Condition 8.2.
38   General Conditions 9.1 and 9.2.

- the provider's name and major office address;

- a description of the Publicly Available Telephone Services offered;

- where the provider renders any subscription charge or periodic rental charge, details of which Publicly Available Telephone Services are included within such charge;

- the provider's standard tariffs, including details of standard discounts and special and targeted tariff schemes, relating to access, all types of usage charge and any maintenance services;

- any compensation and/or refund policy, including specific details of any compensation and/or refund schemes offered;

- any types of maintenance service offered;

- the standard contract conditions offered, including any relevant minimum contractual period; and

- any available dispute resolution mechanisms, including those developed by the provider.

Publication of the information must be effected by sending a copy of this information to end-users upon request and placing the information on the provider's website or at their major offices for inspection free of charge by members of the general public.

3.2.11   *General Condition 11—Metering and Billing*[39]

Under paragraphs 11.1 and 11.2, providers of Public Electronic Communications Services must render accurate bills to end-users, not exceeding the true extent of the service actually provided. Providers must retain such records as may be necessary to comply with this obligation provided that nothing in these paragraphs can oblige the retention of records for more than 15 months.

Providers of Publicly Available Telephone Services whose relevant turnover exceeds £40 million, must, in addition to the general obligation concerning billing and records imposed on Providers of Public Electronic Communications Services, obtain approval of their metering and billing system from a relevant approval body.[40]

3.2.12   *General Condition 12—Itemised bills*

Providers of Publicly Available Telephone Services, except when service is provided on a pre-paid basis or where the subscriber has a free alternative means of adequately monitoring their usage and expenditure, must provide to their subscribers, on request, and either at no extra charge or for a reasonable fee, a basic level of itemised billing. The provider should ensure that each itemised bill shows a sufficient level of detail to allow the subscriber to verify and control the charges incurred and to adequately monitor usage and expenditure and thereby exercise

---

39   Condition 11 was amended in OFCOM, *The Ofcom Metering and Billing Scheme,* 15 July 2008.
40   General Conditions 11.3–11.6. The approval bodies are listed in Condition 11.9(b). See also OFTEL, *Oftel Metering and Billing Direction,* 22 July 2003, as amended in OFCOM, *The Ofcom Metering and Billing Scheme,* 15 July 2008; and OFCOM, *Total Metering and Billing Systems Approval Scheme Guide,* 3rd Edition, 28 June 2005.

a reasonable degree of control over their bills. Free calls and calls to helplines must not be included in the itemised bill. Provision of an itemised bill can be substituted with alternative means, free of charge, to monitor usage and expenditure, such as making available an on-line usage review.

3.2.13   General Condition 13—Non-payment of Bills

Where a subscriber to fixed Publicly Available Telephone Services has not fully paid the bill for such services, any measures taken by the provider to effect payment or disconnection must:

- be proportionate and not unduly discriminatory;

- give due warning to the subscriber beforehand of any consequent service interruption or disconnection; and

- except in cases of fraud, persistent late payment or non-payment, confine any service interruption to the service concerned, as far as technically feasible.

The provider must publish details of measures it may take to effect payment or disconnection by sending a copy of this information to end-users upon request and placing the information on their website.

3.2.14   General Condition 14—Codes of Practice and Dispute Resolution[41]

A person who provides Public Electronic Communications Services to domestic and small business customers must:

- produce a basic code of practice, in plain English, for its domestic and small business customers which sets out at the very least where customers may obtain the information required under General Condition 10;

- establish and maintain procedures that conform with any applicable code of practice for complaints in relation to the provision of their services; and

- implement and comply with a dispute resolution scheme, including any final decision of the dispute resolution body, for the resolution of disputes between providers and their domestic and small business customers. This scheme must be approved by OFCOM.[42]

These providers, where applicable, are also required to establish and maintain, or to comply with, the following specific codes:

- A Code of Practice for Premium Rate Services and 'NTS' calls (which means calls charged at a fixed rate irrespective of geography, such as to numbers beginning 08 or 05).

---

41 General Condition 14 was amended in OFCOM, *Protecting citizens and consumers from mis-selling of fixed-line telecommunications services,* 13 April 2005; OFCOM, *Providing citizens and consumers with improved information about Number Translation Services and Premium Rate Services,* 19 April 2006; OFCOM, *Regulation of VoIP Services,* 29 March 2007; and OFCOM, *Protecting consumers from mis-selling of telecommunications services,* 21 May 2007.

42 OFCOM has published guidelines on the contents of a complaints-handing scheme which it would be able to approve: see *Customer Codes of Practice for handling complaints and resolving disputes,* 24 May 2005.

- A Code of Practice for sales and marketing. This applies to providers of Public Electronic Communications Services who provide fixed-line narrowband[43] call and/or line rental services.

- A Code of Practice on the 'provision by Service Providers of consumer protection information'. This applies to providers of Public Electronic Communication Services that comprise the conveyance of speech, music or sounds. The code is intended to deal with some consumer-protection issues raised by VoIP services. It requires service providers to provide information to its domestic and small business customers on:
  - service reliability;
  - Emergency Calls;
  - the ability to Port Numbers; and
  - other information for domestic and small business customers.

Where service providers are required to produce codes of practice, such codes are required to be drafted in plain, easy to understand English with copies made available on request and free of charge to any domestic and small business customers. General Condition 14 itself includes, as annexes, guidelines on the contents for codes of practice on premium rate services (Annex 1), NTS services (Annex 2) and the sales and marketing of fixed-line services (Annex 3). The Code on the provision of consumer-protection information is in Annex 4.

3.2.15 *General Condition 15—Special Measures for End-users with Disabilities*

Providers of Publicly Available Telephone Services must comply with a number of special obligations regarding end-users with disabilities. These are:

- consulting the OFCOM 'Consumer Panel' to ensure that the requirements and interests of disabled end-users are fully taken into account;

- providing a directory enquiry service free of charge to the visually impaired;

- providing a text-voice relay service with short codes for emergency operations and certain other call recipients for end-users with speech and/or hearing impairments;

- providing a priority fault repair service at no extra cost over the standard charge to any subscriber with disabilities who has a genuine need for an urgent repair;

- ensuring that subscribers who are so disabled that they are dependent on the telephone are able to participate in a scheme to safeguard telephone services to such subscribers; and

- making available large print or brail versions of customer contracts and bills to the visually impaired.

Providers should ensure that the services listed above are widely publicised, taking into consideration the need to disseminate information in appropriate formats and channels for disabled end-users.

---

43 The restriction of this obligation to fixed-line providers only looks increasingly anachronistic as mobile services are increasingly substituted for fixed ones.

3.2.16   *General Condition 16—Provision of Additional Facilities*

Providers of Public Telephone Networks must, subject to technical feasibility and economic viability, provide dual-tone multi-frequency ('DTMF') dialling and calling line identification ('CLI') facilities, unless OFCOM directs that this obligation shall not apply in all or part of the United Kingdom on the basis that there is already sufficient access to these facilities in the relevant areas.[44]

3.2.17   *General Condition 17—Allocation, Adoption and Use of Telephone Numbers*[45]

Providers of Electronic Communications Networks or Services may not adopt numbers from the national telephone numbering plan unless such numbers have been allocated to them or the provider has been authorised to adopt them. The provider is required to have a numbering plan for all its allocated telephone numbers and, except where OFCOM consents in writing, this must conform to the National Numbering Plan.[46]

When applying for an allocation or reservation of telephone numbers, the provider must use the appropriate application form,[47] provide such information as is required and provide any further information that OFCOM may request. OFCOM may withdraw an allocation of telephone numbers where the provider has not adopted the telephone numbers within 6 months or, in relation to a series of telephone numbers, the provider has not adopted them to a serious extent within six months.

Where customers of a provider are making calls to either United Kingdom-wide numbers (03) or Personal Numbering Services (070), the provider shall comply with the designations for those numbers in the National Telephone Plan.[48]

3.2.18   *General Condition 18—Number Portability*[49]

Providers of Electronic Communications Networks or Services are required to

---

44   On CLI, see also OFCOM, *Guidelines for the provision of Calling Line Identification Facilities and other related services over Electronic Communications Networks Version 2*, as amended on 26 April 2007.

45   General Condition 17 was amended in OFCOM, *National Single Non-Emergency Number*, 8 March 2006; OFCOM, *Sexual Entertainment Services (SES) Statement*, 8 March 2007; OFCOM, *Raising confidence in telephone numbers: amending General Condition 17*, 31 May 2007. In OFCOM, *Consumer protection test for telephone number allocation*, 17 May 2007, OFCOM considered further amendments, but in OFCOM, *Consumer protection test for telephone number allocation*, 30 September 2008, decided not to proceed with these. Annex 3 of the latter document includes a set of guidelines relevant to this condition.

46   Available at http://www.ofcom.org.uk/telecoms/ioi/numbers/.

47   These application forms were modified in OFCOM, *Consumer protection test for telephone number allocation*, 30 September 2008.

48   See OFCOM, *Guidance on bodies eligible to use '030' telephone numbers*, Version 2, 31 May 2007.

49   General Condition 18 was amended in OFCOM, *Number Portability and technology neutrality*, 30 March 2006; OFCOM, *Regulation of VoIP Services*, 29 March 2007; and OFCOM, *Arrangement for porting phone numbers when customers switch supplier, a review of General Condition 18*, 17 July 2007. Following the Competition Appeal Tribunal's judgment in *Vodafone v OFCOM*, [2008] CAT 22, the modifications made to General Condition 18 in *Telephone number portability for consumers switching suppliers*, 29 November 2007, have been set aside .

provide Number Portability[50] as soon as it is reasonably practicable, on reasonable terms, including charges, to any of its subscribers who so request it. The provider is also required to provide Portability[51] (other than Paging Portability) as soon as is reasonably practicable in relation to a request by another provider on reasonable terms. In relation to the porting of mobile numbers where the request is for the porting of less than 25 mobile telephone numbers, the numbers are required to be ported within two business days. Any charges for the porting of numbers must be in accordance with this condition.[52]

In *Vodafone v. OFCOM*,[53] Vodafone appealed to the CAT a decision by OFCOM[54] to modify General Condition 18 to require establishment of an 'All Calls Querying of a Common Database' (CDB) (populated initially with ported mobile numbers but to later include ported fixed numbers). The CDB would enable any provider to route calls directly to ported numbers without the need to route the call to the previous network first for onward routing to the recipient. The decision also required that the porting process for mobile numbers would be reduced to two hours, rather than two days, and that only the new provider would need to be contacted by the customer.[55] Vodafone argued that OFCOM had failed to undertake a 'sufficiently rigorous analysis of the costs and benefits of the Decision.' The CAT concluded that OFCOM had not carried out its cost benefit analysis to the requisite standard. In particular, the CAT considered that the cost benefit analysis contained unreliable estimates of the costs of direct routing and relied upon insufficiently justified or explained benefits and was therefore flawed. Having reached this decision, the CAT said that it was not required to reach a decision on two hour number porting.

3.2.19 General Condition 19—Provision of Directory Information

Where a person who provides Electronic Communications Networks or Services has been allocated Telephone Numbers in accordance with General Condition 17, the provider is required to meet all reasonable requests from any person to make available the directory information of its subscribers and any other end-user assigned a number originally allocated to the provider. Where a provider has been authorised to use a telephone number allocated to another person, it is required on request to supply the directory information for the subscribers using those telephone numbers.

50 'Number Portability' is not to be confused with 'Portability.' The first refers to the ability of a PATS customer to change service providers without changing his telephone number. 'Portability,' on the other hand, refers to the obligation on *all* service providers to assist PATS providers with the process of porting customers' numbers.
    Concerning the definition of PATS, see General Condition 18 above.
51 Portability is the facility provided by one provider of electronic communications networks or services to another provider that enables a subscriber to retain his PATS using the same telephone number when changing provider. Note that for the purposes of this General Condition, PATS is specially defined.
52 Concerning portability of VoIP numbers, see OFCOM, *Regulation of VoIP services*, 29 March 2007, Chapter 5.
53 Case No.1094/3/3/08, dated 18 September 2008.
54 OFCOM, *Telephone number portability for consumers switching suppliers*, 29 November 2007.
55 OFCOM's decision also had implications for the system of mobile call termination as under the current system the terminating network operator makes a charge for each call terminated on its network (See Chapter 4, section 4.1.6). Given the current system of onward routing the previous network passes the call termination charge from the originating network to the recipient network after deducting a compensatory charge known as a donor conveyancing charge. Under OFCOM's decision this charge would no longer be payable. See *T-Mobile (UK) Limited v OFCOM* (Case No.1093/3/3/07) concerning an appeal by T Mobile on OFCOM's approach to Donor Conveyancing Charges.

3.2.20   General Condition 20—Non-geographic Numbers

Where a provider of Electronic Communications Networks or Services adopts non-geographic numbers (such as numbers starting 0845), it shall, where technically and economically feasible, ensure that end-users in the European Community are able to access those numbers—although access may be limited if the subscriber so chooses.

3.2.21   General Condition 21—Quality of Service

Providers of Public Electronic Communications Services must, at OFCOM's direction, publish comparable, adequate and up to date information for end-users on the quality of their services. This condition is intended to allow consumers to compare the quality of service of different providers.

3.2.22   General Condition 22—Service Migrations[56]

Providers of broadband services are required, following a request by an end-user, customer or another communications provider to migrate, or where applicable, connect a broadband service to comply with the provisions of the MAC Broadband Migrations Process (attached as Annex 1 to the condition). Where the provisions of the MAC Broadband Migrations Process are not applicable they are instead required to:

- facilitate the migration or connection in a fair and reasonable manner;
- ensure that the migration or connection is carried out within a reasonable period;
- ensure that the migration or connection is carried out with minimal loss of service; and
- assist with and facilitate requests for the migration in instances where the other provider has failed or refused to comply with the MAC Broadband Migrations Process in a manner that is fair and reasonable.

3.3   Procedures for Setting, Modifying and Revoking Conditions

OFCOM must not set or modify a condition unless it is satisfied that the condition or modification is:[57]

- objectively justifiable in relation to the networks, services, facilities, apparatus or directories to which it relates;
- not such as to discriminate unduly against particular persons or against a particular description of persons;

---

56 General Condition 22 was amended in OFCOM, *Broadband migrations: enabling consumer choice,* 13 December 2006.
57 Communications Act 2003, section 47.

- proportionate to what the condition or modification is intended to achieve; and

- in relation to what it is intended to achieve, transparent.

Before setting, modifying or revoking a condition, OFCOM must publish a notification: [58]

- stating that they are proposing to set, modify or revoke the conditions that are specified in the notification;

- setting out the effect of those conditions, modifications or revocations;

- giving their reasons for making the proposal; and

- specifying the period within which representations may be made to OFCOM about their proposal.

That period must end no less than one month after the day of the publication of the notification. [59]

---

58 Communications Act 2003, section 48(2).
59 Communications Act 2003, section 48(3).

Chapter 4

Significant Market Power (SMP) and Other Conditions

Michael H. Ryan and Simon Cloke
Arnold & Porter (UK) LLP*

ARNOLD & PORTER (UK) LLP

In addition to the General Conditions of Entitlement described in Chapter 3, providers of electronic communications network or services or associated facilities may be made subject to other conditions of four types[1]:

- Significant market power ('SMP') conditions

- Universal service conditions

- Access-related conditions

- Privileged supplier conditions.

4.1 SMP Conditions

OFCOM may impose significant market power conditions on a person where a particular market is considered not to be effectively competitive as a result of the fact that that person has a dominant position or the equivalent in the market. A dominant position exists where the person has a position of economic strength affording him the power to behave to an appreciable extent independently of competitors, customers and ultimately consumers[2]. SMP conditions are intended to 'remedy' the effects of such dominance. They may relate to services (see § 4.1.2) or apparatus (see § 4.1.3)[3].

4.1.1 The market analysis procedure

In determining whether a person has SMP in relation to a particular market, and in deciding what conditions should be imposed in order to remedy that situation, OFCOM is required to follow a 'market analysis procedure' defined in the Framework Directive, articles 14–16. The European Commission has adopted

---

*   Solicitors, Tower 42, 25 Old Broad Street, London, EC2N 1HQ, England. Tel: +44 (0) 20 7786 6100.
    e-mail: michael_ryan@aporter.com
1   Communications Act 2003, section 45(2)(b).
2   Communications Act 2003, sections 45(2)(b)(iv) and 78(1) and (2), and the Framework Directive, Directive 2002/21/EC, OJ L 108, 24.04.2002, p. 33, article 14 and Annex II.
3   Communications Act 2003, section 45(7).

Guidelines on Market Analysis and the Assessment of Significant Market Power[4] (the 'Guidelines') and a *Recommendation on relevant product and service markets* (second edition) (the '2007 Recommendation')[5]. OFCOM is required to take account of these in conducting a market analysis[6].

The market analysis procedure is conducted in the three steps summarised in Figure 4.1 and explained below.

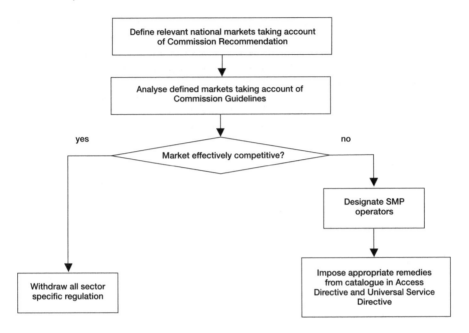

Figure 4.1—Market Analysis Procedure

The first step is market definition. The 2007 Recommendation defines, in accordance with the principles of competition law[7], the product and service markets within the electronic communications sector whose characteristics may be such as to justify the imposition of *ex ante* regulatory obligations. The Recommendation identifies seven such markets—one retail and six wholesale. The 2007 Recommendation supersedes

4   OJ C 165, 11.07.2002, p. 6.
5   Commission Recommendation 2007/879/EC of 17 December 2007 on relevant product and service markets within the electronic communications sector susceptible to ex ante regulation in accordance with Directive 2002/21/EC, C(2007) 5406, OJ L 344, 28.12.2007, p. 65. The Commission is required to update its recommendations on market definition regularly: Framework Directive, article 15(1). The Recommendation replaces the 2003 Recommendation, referred to *infra*.
6   The Communications Act 2003, section 79(3) requires that OFCOM take 'due account' of 'all applicable guidelines and recommendations which (a) have been issued or made by the European Commission ....' The Framework Directive, article 15(3) requires national regulatory authorities to take 'the utmost account' of the Commission's recommendations and guidelines in defining relevant markets.
7   Thus, the relevant product market comprises goods or services which are substitutable from the consumer or the producer perspective (demand/supply-side substitution). The relevant geographic market comprises the area where the conditions of competition are sufficiently homogeneous and which can be distinguished from other areas. Relevant geographic markets may be local, regional, national, or include two or more countries (e.g., pan-European, EEA or global markets): Guidelines, section 1.3.

a similar recommendation issued in 2003 (the '2003 Recommendation')[8]. The 2003 Recommendation identified eighteen markets as susceptible to *ex ante* regulation. Many of the market analyses undertaken under the 2003 Recommendation have yet to be reconsidered since the issuance of the 2007 Recommendation and the former therefore retains its relevance. Figure 4.2 lists the eighteen markets identified in the 2003 Recommendation and shows their relationship to the seven markets identified in the 2007 Recommendation[9].

Step 2: Having defined the relevant markets in Step 1, OFCOM must then proceed to determine whether each of these markets is, or is not, effectively competitive; in other words, whether one or more operators dominate any relevant markets. An operator will be judged dominant if, either individually or jointly with others, it enjoys a position of economic strength affording it the power to behave to an appreciable extent independently of competitors, customers and consumers[10]. While market share is one factor taken into account when assessing the existence of a dominant position, other relevant factors will be taken into account. These include[11]:

- overall size of the undertaking
- control of 'essential facility' type infrastructures
- technological advantages
- absence of countervailing buying power
- economies of scale and scope
- vertical integration
- highly developed distribution and sales network
- absence of potential competition.

When an operator has SMP on a specific market, it may also be deemed to have SMP on a closely related market where the links between those markets are such as to allow the market power held in one market to be leveraged into the second market[12]. OFCOM's SMP findings are summarised in § 4.1.4.

Step 3: If a market is effectively competitive, OFCOM must withdraw any *ex ante* regulatory obligations on an operator that may be in place and may not impose or maintain any new ones.[13] Conversely, if OFCOM finds that a particular market is not effectively competitive because an operator exercises SMP, OFCOM is required to set such SMP conditions as it considers appropriate[14]. The choice of remedies should be justified in light of the objectives of the new framework[15]. It should also

---

8 Commission Recommendation 2003/311/EC of 11 February 2003 on relevant product and service markets within the electronic communications sector susceptible to ex ante regulation in accordance with Directive 2002/21/EC, OJ L 114, 8.5.2003, p. 45.
9 For a discussion of the issues that arise from the transition from the 2003 Recommendation to the current Recommendation, see *Commission Staff Working Document, Explanatory Note, Accompanying document to the Commission Recommendation [etc.]*, second edition, 13 November 2007, SEC(2007) 1483/2, issued with the Recommendation.
10 See Guidelines, section 3.
11 See Guidelines, paragraph 78.
12 Framework Directive, Article 14(3).
13 Framework Directive, Article 16(3); Guidelines, paragraph 113.
14 Framework Directive, Article 16(4); Guidelines, paragraph 113.
15 Guidelines, paragraph 117.

2003 Recommendation	2007 Recommendation
Retail Level	
Market 1 - Access to the public telephone network at a fixed location for residential customers	Market 1 - Access to the public telephone network at a fixed location for residential and non-residential customers
Market 2 - Access to the public telephone network at a fixed location for non-residential customers	
Market 3 - Publicly available local and/or national telephone services provided at a fixed location for residential customers	*Removed*
Market 4 - Publicly available international telephone services provided at a fixed location for residential customers	*Removed*
Market 5 - Publicly available local and/or national telephone services provided at a fixed location for non-residential customers	*Removed*
Market 6 - Publicly available international telephone services provided at a fixed location for non-residential customers	*Removed*
Market 7 - The minimum set of leased lines	*Removed*
Wholesale Level	
Market 8 - Call origination on the public telephone network provided at a fixed location	Market 2 - Call origination on the public telephone network provided at a fixed location
Market 9 - Call termination on individual public telephone networks provided at a fixed location	Market 3 - Call termination on individual public telephone networks provided at a fixed location.
Market 10 - Transit services in the fixed public telephone network	*Removed*
Market 11 - Wholesale unbundled access (including shared access) to metallic loops and sub-loops for the purpose of providing broadband and voice services	Market 4 - Wholesale (physical) network infrastructure access (including shared or fully unbundled access) at a fixed location
Market 12 - Wholesale broadband access	Market 5 - Wholesale broadband access. This market comprises non-physical or virtual network access including 'bit-stream' access at a fixed location
Market 13 - Wholesale terminating segments of leased lines	Market 6 - Wholesale terminating segments of leased lines, irrespective of the technology used to provide leased or dedicated capacity
Market 14 - Wholesale trunk segments of leased lines	*Removed*
Market 15 - Access and call origination on public mobile telephone networks	*Removed*
Market 16 - Voice call termination on individual mobile networks	Market 7 - Voice call termination on individual mobile networks
Market 17 - The wholesale national market for international roaming on public mobile networks	*Removed*
Market 18 - Broadcasting transmission services, to deliver broadcast content to end users	*Removed*

Figure 4.2—Markets Susceptible to ex ante Regulation: 2003 Recommendation and 2007 Recommendation

be proportionate, which implies that it should be the least burdensome option possible to achieve the regulatory aim[16]. The SMP conditions that OFCOM has set in respect of operators found to have SMP are described in §§ 4.1.5 and 4.1.6.

This three step process must be repeated periodically to ensure that regulatory obligations are adapted to the evolution of the market[17].

On 13 November 2007, the European Commission published a wide-ranging Proposal to amend the provisions of the existing regulatory framework that would affect, inter alia, the powers of the NRAs and the Commission in respect of remedies and make certain procedural changes to the market analysis process. More details are given in Chapter 1. At the time of writing, the Commission's Proposal is still being considered by the European Parliament and by Member State governments in the Council.

Figure 4.3—Proposal for Amendment of the 2002 Regulatory Framework

During the market analysis process, the role of the Commission is very important. Not only does it start the procedure by adopting a recommendation on relevant markets, it is required to review all of OFCOM's decisions on market analysis that would affect the trade between member states[18]. It can veto a market definition that differs from those of the Recommendation and an SMP (or a non-SMP) designation[19]; and where it does so, OFCOM is required to withdraw the proposed measure[20]. The Commission may also give a non-binding opinion on other relevant matters, such as the choice of regulatory remedies[21]. The Commission has not exercised a veto in relation to any OFCOM proposal, but it has in several cases expressed an opinion on the proposed choice of remedies[22].

4.1.2 Subject-matter of SMP Services Conditions

The subject-matter of the SMP conditions that OFCOM may impose on an operator found to have significant market power in a particular market include the following. These remedies correspond with the remedies that national regulatory authorities are authorised to impose by the Access Directive and the Universal Service Directive[23].

16 Guidelines, paragraphs 117–118.
17 See Communications Act 2003, section 84.
18 OFCOM is required to notify the European Commission and national regulatory authorities of such measures: Communications Act 2003, section 80(3).
19 The Framework Directive provides that, where an NRA defines markets in a way that does not conform to Commission guidelines and recommendations on market definition (see above) and affects trade between member states, and the Commission considers that giving effect to the proposal would create a barrier in relation to the single European market, or the Commission has serious doubts as to whether the proposal would be compatible with the requirements of any Community obligations, the Commission may require the NRA to withdraw the proposal.
20 See Communications Act 2003, section 82(3).
21 Framework Directive, Article 7(3).
    The Commission's proposals for reform of the 2002 Regulatory Framework released in November 2007 (see Figure 4.3) include a proposed amendment to the Framework Directive which would extend the Commission's veto power to remedies. See Chapter 1 for the latest position at time of writing.
22 The Commission expressed reservations, for example, about OFTEL's conclusion that call termination on 3G networks may not be in the same market as 2G call termination: see Letter from the European Commission to OFTEL, 29 August 2003, SG (2003) D/231466.
23 See the Access Directive, Directive 2002/19/EC, OJ 108, 24.4.2002, p. 7, articles 8–13; and the Universal Service Directive, Directive 2002/22/EC, OJ 108, 24.4.2002, p. 51, articles 17–19.

- A requirement to provide network access[24]. (A typical condition of this type requires a person to provide access to its network on reasonable request at fair and reasonable prices);

- a requirement not to unduly discriminate[25]. (A typical condition of this type requires a person not to unduly discriminate against particular persons or against a particular description of person in relation to network access);

- a requirement to publish information so as to ensure transparency[26]. (A typical condition of this type requires a person to publish information concerning prices or quality of service or to give notice of an intention to change prices);

- a requirement to publish conditions on which a person is prepared to enter into an access contract[27]. (A typical condition of this type requires a person to publish a 'reference interconnection offer');

- a requirement to include certain terms and conditions in an access contract or to modify such terms[28];

- a requirement to maintain separate accounts for different parts of its business, or to use particular accounting methods[29];

- a requirement to comply with price controls, cost recovery, cost orientation of tariffs and cost accounting rules[30]. (A typical condition of this type might impose a price cap, a price floor, or mandate specific prices for a particular service);

- a requirement to provide carrier selection and pre-selection services (enabling end-users connected to a specific network to select, either on each separate occasion or else automatically in advance, which service provider they wish to use over that network)[31];

- a requirement to provide leased lines identified by the European Commission under the Universal Service Directive on non-discriminatory and cost-oriented terms; and [32]

- NV additional condition if justified by exceptional circumstances and approved by the European Commission[33].

---

24 Communications Act 2003, section 87(3) and (5). In determining what conditions to set in a particular case, OFCOM must have regard to the factors identified in section 87(4).
25 Communications Act 2003, section 87(6)(a).
26 Communications Act 2003, section 87(6)(b).
27 Communications Act 2003, section 87(6)(c).
28 Communications Act 2003, section 87(6)(d), (e). 'Access contract' is defined in section 87(12) as a contract under which one person grants another access to their network or to their associated facilities.
29 Communications Act 2003, section 87(7).
30 Communications Act 2003, section 87(9). Price controls, cost recovery, cost orientation of tariffs and cost accounting rules may only be imposed where there is a risk of adverse effects arising from price distortion in the relevant market and the setting of such conditions is appropriate for promoting efficiency and sustainable competition and conferring the greatest possible benefits on end-users: ibid., section 88.
31 Communications Act 2003, section 90.
32 Communications Act 2003, section 92.
33 Communications Act 2003, section 89.
    The Commission's proposals for reform of the 2002 Regulatory Framework released in November 2003 (see Figure 4.3) include a proposed amendment to the Framework Directive which would authorise an NRA to require vertically integrated undertakings to place activities related to the wholesale provision of access products in an independently operating business unit (i.e., to impose 'functional separation') as a remedy where other remedies have failed and would persistently fail to achieve effective competition and there are important and persisting competition problems/market failures identified in several product markets. Commission approval would be required before such a remedy could be imposed (article 13a). See Chapter 1.
    BT has already undergone a form of functional separation on a voluntary basis: see Chapter 5.

Significant Market Power (SMP) and Other Conditions

While the subject-matter of conditions must conform to those authorised by the Communications Act, the specific content of those conditions may differ from market to market.

In addition to these conditions, OFCOM has imposed conditions on SMP providers relating to:

- provision of interconnection circuits[34];

- product management, policy and planning ('PPP')[35].

OFCOM's authority to regulate interconnection circuits derives from their status as a 'technical area' in which OFCOM can apply remedies as part of the overall solution to address SMP in relevant economic markets[36]. Authority to regulate PPP derives from its status as a component of the services in which BT has SMP[37].

SMP should be addressed by the imposition of regulation at the wholesale level in preference to measures imposed at the retail level. That is the effect of section 91 of the Communications Act 2003, which provides that, where a person is found to have SMP in a retail market for electronic communications services, and OFCOM is of the view that the imposition of access-related conditions or SMP conditions in relation to the underlying wholesale market would not be sufficient to satisfy their statutory duties[38], OFCOM may impose conditions on that person relating to the relevant retail market as appropriate[39].

4.1.3   Subject-matter of SMP Apparatus Conditions

OFCOM may impose SMP conditions relating to the following subjects on a supplier found to have SMP in an apparatus market[40]:

---

34   See Figure 4.4 and OFCOM, *Explanatory Statement and Notification of decisions on BT's SMP status and charge controls in narrowband wholesale markets*, Annex 5, paragraphs A5.90 et seq. for an example of a situation where interconnection circuits are regulated. The determinations made by OFCOM resulted in Condition AA being imposed on BT.

35   Condition PA imposed on BT is an example of a situation where PPP has been imposed as a remedy: see Figure 4.4. See OFCOM, *Review of BT's product management, policy and planning (PPP) charge*, 30 July 2004, paragraphs 1.1 to 1.3 and 9.1 to 9.2 for a definition of PPP.

36   See Access Directive, article 5(1)(a). See also OFCOM, *Review of BT's Network Charge Controls*, 18 August 2005, paragraph 2.9.

  In the Explanatory Memorandum to its 2003 Recommendation, the European Commission says that:

   '[i]n dealing with lack of effective competition in an identified market, it may be necessary to impose several obligations to achieve an overall solution. For instance, it may often be the case that adjacent or related remedies are applied to technical areas as part of the overall obligation that addresses SMP on the analysed market. If specific remedies are thought to be necessary in a specific narrow technical area, it is not necessary or appropriate to identify each technical area as a relevant market in order to place obligations in that area. An example would be where an obligation to provide unbundled access to the local loop is complemented by related obligations concerning access to co-location facilities.'

  See also section 4.3 below under technical and operational matters related to SMP conditions.

37   See OFCOM, *Review of BT's Network Charge Controls*, 18 August 2005, paragraph 2.10.

38   These duties are defined in the Communications Act 2003, section 4.

39   Communications Act 2003, section 91.

40   Communications Act 2003, section 93.

- a requirement to maintain separate accounts and to use particular accounting methods and cost accounting systems for that purpose[41];

- a requirement to comply with price controls in the case of hiring of telephones which are hardwired to a network[42].

4.1.4 SMP Findings

OFCOM has determined that BT has SMP in the retail services markets in the UK (except Hull) identified in Figure 4.4 and in the wholesale services markets in the UK (generally except Hull) identified in Figure 4.5. OFCOM has also determined that KCOM (the entity providing local telephone service in Kingston-on-Hull) has SMP in various retail and wholesale services markets in Hull.

OFCOM has determined that each provider of public electronic communications networks (including mobile operators) has SMP in respect of geographic call termination on its own network[43].

OFCOM has determined that ntl:broadcasting and Crown Castle and NTL each have SMP in the area served by their respective masts and sites in respect of digital and analogue terrestrial television transmission[44].

In recent years, OFCOM has focused regulation upstream, on wholesale rather than retail markets. As a result, no retail leased line markets—other than the market for traditional interface low bandwidth service[45]—are subject to regulation. Furthermore, all remaining retail price controls imposed on BT were allowed to lapse in July 2006[46]. (Obligations not to unduly discriminate and to publish charges continue to apply to the provision of exchange lines and most categories of retail call)[47].

Some wholesale markets have also been 'deregulated.' For example, the following wholesale markets have been found to be competitive and, as a result, no SMP conditions have been imposed:

- wholesale mobile access and call origination[48];

- wholesale international services[49];

---

41 Communications Act 2003, section 93(2).
42 Communications Act 2003, section 93(3).
43 OFCOM, *Mobile call termination*, 27 March 2007.
44 OFCOM, *Broadcasting transmission services: a review of the market*, 11 November 2004.
45 OFCOM, *Business Connectivity Market Review*, 8 December 2008.
46 OFCOM, *Retail Price Controls*, 19 July 2006. The elimination of BT retail price controls was linked to BT's September 2005 agreement with OFCOM to undertake to provide various wholesale products on a 'equivalence of inputs' (EoI) basis (an arrangement which included creation of an operationally separate access division now known as 'OpenReach'). See *Retail Price Controls, op. cit.*, paragraph 1.10.
47 However, at the time of writing, OFCOM is consulting on removing the SMP designations from narrowband retail markets—see Note 6 to figure 4.4 in this chapter.
48 OFCOM, *Mobile access and call origination services market*, 4 August 2003. As a consequence of this finding, there is no *ex ante* requirement on UK mobile network operators (MNOs) to give MVNOs (mobile virtual network operators) access on pre-defined terms. Not all Member States in the EU have reached the same conclusion—though the exclusion of this market (market 15) from the 2007 Recommendation means that any such *ex ante* regulation in other Member States is likely to be withdrawn.
49 OFCOM, *Review of Wholesale international services markets*, 7 July 2006.

Figure 4.4—BT SMP conditions (retail markets)

Remedies	Fixed PSTN Access	Analogue exchange line services	ISDN2 exchange line services	Call origination on the PSTN	Local Calls	National Calls	International Calls - "Category A"	International Calls - "Category B"	Calls to Mobiles
Other									
Cost Orientation (Basis of charges)									
Publication of reference offer									
Continued supply of leased lines currently being provided									
Publication of delivery and repair times.									
Publication of charges and notification of changes		✓	✓		✓	✓	✓	✓	✓
Price controls									
No undue discrimination		✓	✓		✓	✓	✓	✓	✓

R = Residential
NR = Non-Residential

Market (UK excl Hull unless indicated otherwise)

Market No. in Commission's 2003 [2007] Recommendation

| OFCOM Condition No. | D | | | D | | | | | |

Operator-assisted Calls			✓	✓	
Leased Lines					
Low bandwidth traditional interface services ("8 Mbits/s)	✓			✓	Note 5
High bandwidth traditional interface services (8–45 Mbits/s)					
Very high bandwidth traditional interface services (45–155 Mbits/s)					
Very high bandwidth traditional interface services (> 155 Mbits/s)					Note 2
Low bandwidth alternative interface services ("1Gbit/s)					
High bandwidth alternative interface services (> 1Gbit/s)					

Note 1: The retail price control applicable to this service lapsed on 1 August 2006: see *Retail Price Controls*, 19 July 2006. Some price constraints continue to apply in certain cases under the Universal Services Condition.

Note 2: Finding of no SMP in these markets: see OFCOM, *Business Connectivity Market Review*, 8 December 2008.

Note 3: "Category A" routes are those 123 routes that were found in 2003 to be competitive at wholesale level. Retail call origination to those routes is treated as one single retail market. Category B" routes are those 112 routes that were found in 2003 to be not competitive at wholesale level. Retail call origination on those routes is treated as a separate retail market in respect of each of those routes.

Note 4: The obligation to notify charges does not apply to prices which remain above a price floor in the case of customers whose expenditure exceeds £1m p.a.: see *Replicability: the regulation of BT's retail business exchange line services*, 29 May 2007.

Note 5: Obligation to provide: BT is required to supply existing and new 2 Mbit/s retail low bandwidth leased lines to third parties on reasonable request. In addition, BT has made the following voluntary undertakings: (1) To continue to supply new analogue retail circuits until 1 January 2011 or earlier if, subject to industry agreement and consent by OFCOM, the underlying platform is closed at an earlier date; and (2) To continue to supply new sub-2Mbit/s retail circuits until 1 January 2011 or earlier if, subject to industry agreement and consent by OFCOM, the underlying wholesale products are withdrawn from new supply at an earlier date.

Obligation not to unduly discriminate: In respect of all analogue and digital services at speeds up to and including 8 Mbit/s, BT is required not to unduly discriminate.

Obligation to publish a reference offer: In respect of all analogue and digital services of speed up to and including 2 Mbit/s, BT is required to publish prices, terms and conditions, and to notify on the same day of entering into force any changes to those prices terms and conditions. In addition, BT has made the following voluntary undertakings: (1) Not to increase its prices for analogue services more quickly than the rate of inflation (RPI-0%) for a period two years following the publication of the Business Connectivity Market Review Statement i.e. from 2008 to 2010; and (2) To comply with a further two-year cap, the level of which would be agreed with OFCOM prior to 2011. If BT fails to adhere to its pricing commitment, or if BT and OFCOM should fail to reach agreement on the two-year cap for 2011–12, a cost orientation condition in relation to the price of analogue services would then come into effect. See *Business Connectivity Market Review*, 8 December 2008.

Note 6: In *Fixed Narrowband Retail Services Markets*, 19 March 2009, OFCOM initiated a consultation on the future of regulation of fixed narrowband retail services. OFCOM proposed that BT no longer has SMP in:
• retail fixed narrowband telephone lines for business and residential consumers
• retail fixed narrowband calls for business and residential consumers
and proposed to remove all relevant SMP conditions. OFCOM also indicated that, while it considered that BT still has SMP in ISDN2 and ISDN30 lines, in its view that the current retail remedies are no longer effective and are even potentially counterproductive in the development of enhanced competition and proposed to remove all relevant retail SMP conditions and rely solely on wholesale remedies. See Figure 4.5, Note 17.
OFCOM is consulting on these proposals and, as of 31 May 2009, no determination had been published.

Figure 4.5—BT SMP conditions (wholesale markets)

80

Market (UK excl Hull unless indicated otherwise) — R = Residential, NR = Non-Residential

OFCOM Condition No.	Market No. in Commission's 2003 [2007] Recommendation	Market / Service	No undue discrimination	Charge controls	Provision of quality of service information	Publication of reference offer	Notification of charges	Notification of technical information	Provision of network access on reasonable request	Provision of new network access	Cost orientation (Basis of charges)	Accounting Separation	Other
AA		**Exchange line services**											
		Analogue exchange line services	✓		✓	✓	✓	✓	✓	✓	✓		Notes 1 & 17
		ISDN2 exchange line services	✓		✓	✓	✓	✓	✓	✓	✓		
		ISDN30 exchange line services	✓		✓	✓	✓	✓	✓	✓			
AA	8	**Call origination on the PSTN**											
		Call origination on the PSTN	✓	✓	✓	✓	✓	✓	✓	✓	✓		Notes 2, 16 & 17
AA	10	**Transit services on the PSTN**											
		Local tandem conveyance and transit	✓	✓	✓	✓	✓	✓	✓	✓	✓		Notes 3, 16 & 17
		Inter-tandem conveyance and transit							✓	✓			Notes 4 & 17
		Single transit	✓	✓	✓	✓	✓	✓	✓	✓	✓		Note 17
KA		**International call conveyance**											
		International call conveyance											Note 5

										Notes
		Call termination on fixed networks								
BA	9	Fixed geographic call termination	✓	✓	✓	✓	✓	✓	✓	Notes 16 & 17
		Narrowband internet termination								
		Unmetered narrowband internet termination								Note 6
		Wholesale local access								
FA	11	Wholesale local access	✓	✓	✓	✓	✓	✓	✓	Note 7
		Wholesale asymmetric broadband access								
EA		Market 1	✓	✓	✓	✓	✓	✓		
	12	Market 2	✓	✓	✓	✓	✓	✓		
		Market 3								Note 13
		Wholesale symmetric broadband origination								
G		Traditional interface symmetric broadband origination (TISBO) ("8Mbits/s)	✓	✓	✓	✓	✓	✓	✓	Notes 8 & 14
GG		TISBO (8–45Mbits/s) (excl Central and East London)	✓	✓	✓	✓	✓	✓	✓	
GH	[6]	TISBO (45–155Mbits/s) (excl Central and East London)	✓	✓	✓	✓	✓	✓	✓	
HH		Alternative interface symmetric broadband origination (AISBO) ("1Gbits/s)	✓	✓	✓	✓	✓	✓	✓	
		TISBO (>155Mbits/s)								
		Alternative interface symmetric broadband origination (AISBO) (> 1Gbits/s)								Note 9

Figure 4.5—*continued*

OFCOM Condition No.	Market No. in Commission's 2003 [2007] Recommendation	Market (UK excl Hull unless indicated otherwise)	R = Residential NR = Non-Residential	No undue discrimination	Charge controls	Provision of quality of service information	Publication of reference offer	Notification of charges	Notification of technical information	Provision of network access on reasonable request	Provision of new network access	Cost orientation (Basis of charges)	Accounting Separation	Other
		Wholesale trunk segments												
H	14	Wholesale trunk segments at all bandwidths		✓	✓	✓	✓	✓	✓	✓	✓	✓		Notes 14 & 15
		NTS call termination												
L		NTS call termination												Note 10
		Other												
OA	17	International roaming												Note 11
OA		Imposes a regime for financial reporting and cost accounting across the range of services that are subject to SMP conditions			✓									Note 12
PA		Product management, policy and planning (PPP) and Interconnection circuits												Note 16

Note 1: In response to the findings of SMP in the analogue exchange line residential and business markets, the ISDN2 business market and the ISDN30 business market, BT has been required as one of the remedies imposed by Condition AA to introduce a wholesale line rental product (WLR). See *Review of the Fixed Narrowband Wholesale Exchange Line, Call Origination, Conveyance and Transit Markets*, 28 November 2003, chapter 7. Although services subject to this condition are not generally subject to price controls, prices for WLR are subject to controls: see *Wholesale Line Rental: Reviewing and setting charge ceilings for WLR services*, Annex 1, 24 January 2006. These prices are now under review: See OFCOM, *A New Pricing Framework for Openreach*, 5 December 2008.

Note 2: In response to the finding of SMP in the call origination market, BT has been required as one of the remedies imposed by Condition AA to introduce (i) Carrier Selection (i.e., Indirect Access), (ii) Carrier Pre-Selection (CPS) and (iii) Number Translation Services (NTS) Call Origination. See *Review of the Fixed Narrowband Wholesale Exchange Line, Call Origination, Conveyance and Transit Markets*, 28 November 2003, chapters 6, 7 and 8, respectively.

Note 3: In response to the findings of SMP in the call origination market and the local-tandem conveyance and transit market, BT has been required as one of the remedies imposed by Condition AA to introduce flat rate internet access call origination (FRIACO). See *Review of the Fixed Narrowband Wholesale Exchange Line, Call Origination, Conveyance and Transit Markets*, 28 November 2003, chapter 5.

Note 4: Finding of SMP revoked: see *Review of BT's Network Charge Controls*, 18 August 2005.

Note 5: Finding of SMP revoked: see *Review of Wholesale international services markets*, 7 July 2006.

Note 6: Finding that BT does not have SMP: see *Wholesale Unmetered Narrowband Internet Termination Services, UK excluding Hull area market*, 28 November 2003.

Note 7: Condition FA also includes a requirement to provide local loop unbundling services on "fair and reasonable terms, conditions and charges" and as OFCOM may direct from time to time. Although services subject to this condition are not generally subject to price controls, prices for certain LLU-related services are subject to controls: see *Review of the wholesale local access market*, 16 December 2004, Annex 2; and *Local loop unbundling: setting the fully unbundled rental charge ceiling and minor amendments to SMP conditions FA6 and FB6*, Annex 1, 30 November 2005. These prices are now under review: See OFCOM, *A New Pricing Framework for Openreach*, 5 December 2008.

Note 8: See also the Directions relating to the provision of access on reasonable request set out in *Business Connectivity Market Review*, 8 December 2008, Schedules 11–15.

Note 9: Finding that BT does not have SMP: see *Business Connectivity Market Review*, 8 December 2008.

Note 10: OFCOM has commenced a review of the market for number translation services and published proposed SMP conditions: *NTS call termination market review*, 22 October 2004. That review has not concluded.

Note 11: OFCOM undertook preliminary work in assessing the wholesale international roaming market on public mobile networks, but that work was discontinued as a result of the adoption of the Roaming Regulation. See § 4.1.7.

Note 12: The regulatory and financial reporting obligations imposed on BT in relation to wholesale international call conveyance "Category B" were removed: see *Review of Wholesale international services markets*, 7 July 2006.

Note 13: Finding that no person has SMP in relation to Market 3: *Review of the Wholesale Broadband Access Markets*, 21 May 2008.

Note 14: Charge controls imposed under previous determinations continue pending the outcome of a consultation on a new potential charge control: *Business Connectivity Market Review*, 8 December 2008, para. 8.60. For the consultation document, see OFCOM, *Leased Lines Charge Controls*, 8 December 2008. OFCOM is consulting on these proposals and, as of 31 May 2009, no determination had been published.

Note 15: BT has SMP in the entire UK market, incl. Hull: see *Business Connectivity Market Review*, 8 December 2008.

Note 16: These network charge controls (NCCs) lapse in September 2009. In *Review of BT Network charge controls*, 19 March 2009, OFCOM initiated a consultation on the future of the controls. OFCOM proposed to identify BT has possessing SMP, and the need for NCCs as a remedy, in the following markets:

- wholesale call origination
- wholesale call termination.

In addition, OFCOM proposed to continue regulation of interconnection circuits and product management, policy and planning (PPP).

OFCOM is not proposing to continue to regulate NCCs in respect of the following services:

- single transit
- local-tandem conveyance
- Flat Rate Internet Access Call Origination (FRIACO).

OFCOM is consulting on these proposals and, as of 31 May 2009, no determination had been published.

Note 17: In *Review of the fixed narrowband services wholesale markets*, 19 March 2009, OFCOM initiated a consultation on the future of regulation of fixed narrowband wholesale services. OFCOM proposed that BT retains SMP in the markets for exchange lines, call origination and call termination in the UK outside Hull, but no longer has SMP in local-tandem conveyance and transit. It also proposed that (1) the inter-tandem conveyance, inter-tandem transit and single transit products form a single market and that BT does not have SMP in this market; and (2) all providers of fixed geographic call termination retain SMP in call termination on their own networks. In markets where BT retains SMP, OFCOM proposed to take the following actions:

- to remove the requirement for BT to provide certain exchange line services (WLR) and call origination services (CPS) in compliance with functional specifications as directed by OFCOM
- to impose additional obligations in relation to ISDN30 exchange lines
- to consider whether it is appropriate to reduce notification periods for price changes, and
- to narrow reporting obligations concerning compliance with SMP obligations.

OFCOM is consulting on these proposals and, as of 31 May 2009, no determination had been published.

83

- wholesale unmetered narrowband internet termination for traffic originating in all parts of the UK except Hull[50];

- unmetered narrowband Internet termination[51];

- inter-tandem conveyance and transit[52];

- alternative interface symmetric broadband origination (AISBO) (>1Gbits/s)[53];

- wholesale broadband access, Market 3; and[54]

- traditional interface symmetric broadband origination (TISBO) (>155Mbits/s)[55].

A feature of recent decisions has been OFCOM's readiness to define different geographic markets within the UK and impose different SMP conditions in respect of those different markets. BT has, for example, decided to remove all regulation in respect of wholesale broadband access in the region it has defined as 'Market 3,' while continuing to impose SMP conditions on BT in respect of Markets 1 and 2.[56] Similarly, OFCOM has removed all regulation in respect of traditional interface symmetric broadband origination at speeds in excess of 8 mbits/s in central and eastern areas of London, whilst retaining regulation elsewhere.[57]

4.1.5   SMP Conditions Imposed on BT

Figures 4.4 and 4.5 identify the types of conditions imposed on BT in respect of retail markets and wholesale markets, respectively, where BT has been found to have SMP.

4.1.6   SMP Conditions Imposed on Other Operators

OFCOM has imposed an SMP condition on all providers of public electronic communications fixed networks requiring them to provide network access upon request, on fair and reasonable terms[58].

OFCOM has imposed SMP conditions on KCOM in respect of retail and wholesale services provided by KCOM in Hull that are similar (though not in every case identical) to those imposed on BT in respect of its corresponding services (see § 4.1.5).

OFCOM has imposed SMP conditions on H3G, O2, Orange, T-Mobile and Vodafone in relation to termination of both 2G and 3G wholesale mobile voice calls. These conditions include:

---

50 OFTEL, *Review of the Wholesale Unmetered Narrowband Internet Termination Market,* 17 March 2003.
51 OFTEL, *Wholesale Unmetered Narrowband Internet Termination Services, UK excluding Hull area market,* 28 November 2003.
52 OFCOM, *Review of BT's Network Charge Controls,* 18 August 2005.
53 OFCOM, *Business Connectivity Market Review,* 8 December 2008.
54 OFCOM, *Review of the Wholesale Broadband Access Markets,* 21 May 2008.
55 OFCOM, *Business Connectivity Market Review,* 8 December 2008.
56 OFCOM, *Review of the Wholesale Broadband Access Markets,* 21 May 2008.
57 OFCOM, *Business Connectivity Market Review,* 8 December 2008.
58 OFTEL, *Review of fixed geographic call termination markets,* 28 November 2003.

- a requirement to provide network access upon request, on fair and reasonable terms;

- a requirement not to unduly discriminate in relation to network access;

- a control applicable to fixed-to-mobile interconnection charges;

- a control applicable to mobile-to-mobile interconnection charges; and

- a requirement to publish charges[59].

OFCOM has imposed SMP conditions on NTL:Broadcasting and Crown Castle in relation to the provision of access to the mast, site network and antennae systems used by them for the purpose of providing analogue and/or digital terrestrial broadcasting transmission services in the U.K[60]. These conditions include:

- a requirement to provide network access upon request, on fair and reasonable terms;

- a requirement not to unduly discriminate in relation to network access;

- a requirement that charges are cost orientated; and

- a requirement to publish a 'reference offer' for network access.

4.1.7 The Roaming Regulation

OFCOM commenced preliminary work on a review of the wholesale international roaming market with a view to determining whether SMP conditions should be imposed, but that work was discontinued as a result of the adoption of the EU 'Roaming Regulation' in 2007 (subsequently amended in 2009)[61]. The Roaming Regulation caps wholesale charges for Community-wide roaming calls, SMS messages, and 'regulated data roaming services' (i.e., data roaming services other than SMS messaging).[62]

---

59   OFCOM, *Mobile call termination*, 27 March 2007. A provision requiring the operators to monitor compliance with the fixed-to-mobile interconnection charges and mobile to mobile interconnection charges was notified under OFCOM, *Monitoring compliance with charge controls,* 18 December 2007, but was subsequently removed by a notification on 11 July 2008.
      Mobile termination has been, and remains, a particularly controversial area of telecoms regulation in the UK and it has been the subject of a number of legal challenges. H3G, in particular, has fought and lost a number of legal proceedings against the finding that it has SMP. In *British Telecommunications Plc v. OFCOM* (case no. 1085/3/07) and *Hutchinson 3G UK Limited v OFCOM* (case no. 1083/3/07) the Competition Appeals Tribunal (CAT) concluded that the charges for connecting to the O2, Orange, T-Mobile and Vodafone networks should be reduced to 4.0 pence per minute (ppm) by 2010/11. (OFCOM had decided that they should fall to 5.1ppm by 2010/11.) The CAT also determined that the charge for connecting to the H3G network should be reduced to 4.4ppm by 2010/11 (1.5ppm less than the price control established by OFCOM). See also *Hutchison 3G (UK) Limited v Office of Communications* (case no. 1047/3/3/04), *infra,* footnote [96].
60   OFCOM, *Broadcasting transmission services: a review of the market,* 28 April 2005.
61   Regulation (EC) No. 717/2007 of 20 June 2007 on roaming on public mobile telephone networks within the Community and amending Directive 2002/21/EC, OJ L 171, p. 32, 29.6.2007, as amended with effect from 30 June 2009 by Regulation (EC) No 544/2009 of 18 June 2009, OJ L 167, p. 12, 29.6.2009.
      In 2007, the European Commission sent statements of objection to mobile network operators O2 and Vodafone alleging that the rates they charged for wholesale international wholesale roaming were unfair and excessive under Community competition laws. See IP/04/994, 26 July 2007.
62   Roaming Regulation, articles 3, 4a and 6a(4). These provisions apply to wholesale charges for 'regulated roaming calls' and 'regulated roaming SMS messages' and 'regulated data roaming services,' respectively, imposed by the operator of a 'visited network' on the operator of a retail customer's 'home network.' The relevant terms are defined in article 2.

The Roaming Regulation also imposes controls on retail pricing. Providers are obliged to introduce per second billing after the first 30 seconds of a call[63] and to introduce optional 'Eurotariffs' and 'Euro SMS-tariffs' with maximum retail charges for Community-wide roaming calls and SMS messages[64].

A customer's home provider is required to alert a roaming customer that he will be subject to roaming charges when making or receiving a call or when sending an SMS message[65]. Home providers must also ensure that their roaming customers are kept informed of charges for the use of regulated data roaming services[66].

By 1 March 2010, home providers must make available arrangements which allow customers to adopt maximum financial limits for use of regulated data roaming services for specified periods of use, after which the provider must cease to provide the service unless and until the customer requests otherwise[67].

The Roaming Regulation expires on 30 June 2012[68].

Practical Issue: Roaming Regulation

In *R. (Telefonica, O2 Europe and others) v. Secretary of State for Business and Regulatory Reform,*[69] UK mobile operators challenged the validity of the Roaming Regulation and the UK implementing measure—the Mobile Roaming (European Communities) Regulations 2007.[70] In the context of that proceeding, the High Court has referred the following questions to the European Court of Justice for determination:[71]

- Is Regulation (EC) No. 717/2007 invalid, in whole or in part, by reason of the inadequacy of Article 95 EC as a legal basis?

- Is Article 4 of Regulation (EC) No. 717/2007 (together with Articles 2(a) and 6(3) insofar as they refer to the Eurotariff and obligations relating to the Eurotariff) invalid on the grounds that the imposition of a price ceiling in respect of retail roaming charges infringes the principle of proportionality and/or subsidiarity?

4.2 Universal Service Conditions

The Communications Act 2003 requires the Secretary of State to designate by order the types of services that must be provided in the UK to secure compliance with the UK's obligations in respect of the provision of universal service under the Universal

---

63 Roaming Regulation, article 3.
64 Roaming Regulation, articles 4 and 4b. These provisions apply to 'regulated roaming calls' and 'regulated roaming SMS messages' imposed by 'home providers,' as defined in article 2. 'Eurotariff' and 'Euro SMS-tariff' are also defined terms.
65 Roaming Regulation, article 6(1).
66 Roaming Regulation, article 6a(1).
67 Roaming Regulation, article 6a(3).
68 Roaming Regulation, article 13.
69 [2007] EWHC 3018 (Admin).
70 SI 2007/1933.
71 Case C-58/08.

Service Directive (the 'Universal Service Order')[72]. The following services have been designated as universal services:[73]

■ connection at a fixed location to the public telephone network;

■ at least one comprehensive directory of all subscribers of publicly available telephone services;

■ at least one comprehensive telephone directory enquiry facility including all subscribers of publicly available telephone services;

■ public pay telephones to meet the reasonable needs of end-users in terms of geographical coverage and the number of telephones, including the ability to make emergency calls free of charge;

■ appropriate tariff options and packages for subscribers on low incomes or with special social needs;

■ facilities and billing and payment methods that enable subscribers to monitor and control their expenditure; and

■ certain special measures for end-users with a disability.

OFCOM is empowered to designate the persons to whom universal service obligations apply,[74] and to set universal service conditions to secure compliance with the obligations set out in the Universal Service Order[75]. All of the above universal services are to be offered by the persons so-designated at an affordable and uniform price throughout the country (unless there is a clear justification for not doing so)[76].

Pursuant to these powers, OFCOM has designated BT and KCOM as universal service providers. BT's universal service obligation extends to the whole of the UK except Hull. KCOM's universal service obligation covers the Hull area only[77].

4.2.1 Universal Service Conditions Imposed on BT

OFCOM has imposed universal service conditions on BT and KCOM. Figure 4.6 summarises the universal service conditions imposed on BT. Those imposed on KCOM are similar.

---

72 Communications Act 2003, section 65. The Electronic Communications (Universal Service) Regulations 2003, SI 33/2003, defines 'universal service' with reference to the obligations defined in articles 4, 5, 6, 7 and 9(2) of the Universal Service Directive.
73 The Electronic Communications (Universal Service) Order 2003, SI 1904/2003.
74 Communications Act 2003, section 66(1); The Electronic Communications (Universal Service) Regulations 2003, SI 33/2003, Regulation 4. OFCOM must carry out this function in accordance with the policy objectives and regulatory principles in Article 8 of the Framework Directive: The Electronic Communications (Universal Service) Regulations 2003, SI 33/2003, Regulation 3(1).
75 Communications Act 2003, section 67(1).
76 The Electronic Communications (Universal Service) Order 2003, SI 1904/2003, s.4.
77 See OFTEL, *Designation of BT and Kingston as universal service providers, and the specific universal service conditions*, 22 July 2003. The Conditions applicable to BT are set out in Part 2. The Conditions applicable to KCOM are set out in Part 3. Part 1 is common to both sets of Conditions.

Figure 4.6—BT universal service conditions

Condition		Summary of Condition
1	Provision of telephony services on request	BT must meet all reasonable requests for Telephony Services,[1] including the ability to send and receive faxes and data at data rates that are sufficient to permit "functional internet access,"[2] at uniform prices.[3]
2	Schemes for consumers with special social needs	BT must make available one or more schemes to assist Consumers who have difficulty affording telephone services, including, in particular, Consumers on low incomes or with special needs. OFCOM may makes directions specifying the requirements to be met by such schemes and the criteria to be applied in deciding eligibility.[4]
3	Provision of call box services[5]	BT must ensure adequate provision of Public Call Boxes and Call Box Services.[6] It must conform with OFCOM directions[7] in removing and re-siting Public Call Boxes and in considering requests for new Public Call Boxes.[8]
4	Provision of relay service for textphone users	BT must fund, within prescribed limits, the operation of a relay service for all End-Users of PATS who need to use Textphones because of their disabilities and enter into an agreement with a relay service provider for that purpose.
5	Tariffs for universal services	The terms on which universal services or facilities are provided must not require payment for unnecessary additional service.
6	Itemised Billing	BT must provide itemised billing at no extra charge. OFCOM may make directions specifying the level of itemisation to be provided.
7	Maintenance and supply of a directory information database and directories[9]	BT must maintain a database for all Subscribers and make the information available to other Communications Providers and providers of Directory Enquiry Facilities and Directories. BT must not discriminate in the treatment of data or information supplied to it, and must have due regard to any Subscriber who has expressed opposition to inclusion of Directory Information about that Subscriber.
8	Quality of Service	BT must regularly publish information concerning its compliance with Conditions 1, 2, 3, 4 and 7 in accordance with the measurement methods defined in the Universal Service Directive, Annex 3, and in any additional standards or requirements directed by OFCOM.

1   OFCOM has published guidance on BT's obligation to provide a connection to the fixed network upon reasonable request: see OFCOM, *Review of the Universal Service Obligation,* 14 March 2006, Annex 6. See OFCOM, *Ebbsfleet Fibre to the Home Pilot,* 1 August 2008 in which OFCOM provided BT with a consent not to apply universal service condition 1.2 in respect of connection and rental charge for telephony services for the Ebbsfleet Pilot subject to a number of provisions.

2   OFCOM has published guidance on what constitutes "Functional Internet Access." The latest is OFCOM, *Review of the Universal Service Obligation,* 14 March 2006, Annex 7. This says that a minimum speed of 28.8 kbps is required.

3   The obligation on BT to provide the services mentioned in this Condition on the basis of uniform prices does not apply where the provision of those services costs BT more than £3400 excluding VAT: see OFCOM, *Review of the Universal Service Obligation,* 14 March 2006, Annex 5.

4   Concerning the eligibility criteria for the schemes currently run by BT, see OFCOM, *Review of the Universal Service Obligation,* 14 March 2006, p. 8.

5   The definition of "Site" in the universal services conditions imposed on BT in "*Designation of BT and Kingston as universal service providers, and the specific universal service conditions*," *supra,* which impacts BT's obligations in relation to the provision of Public Call Boxes referred to in this Condition, was amended in OFCOM, *Review of the Universal Service Obligation,* 14 March 2006, Annex 2.

6   See OFCOM, *Public Call Boxes,* 4 July 2007 in which OFCOM provided BT with consent for non-uniform geographic charging.

7   OFCOM has issued a Direction on removal of Public Call Boxes: see OFCOM, *Review of the Universal Service Obligation,* 14 March 2006, Annex 3.

8   OFCOM has published guidance on procedures for the complete removal of public call boxes and/or call box services from a site: see OFCOM, *Review of the Universal Service Obligation,* 14 March 2006, Annex 4.

9   See OFCOM, *Telephone directory information obligations and regulations*, 10 March 2008 in which OFCOM proposes the revocation of universal service condition 7.

4.2.2 Funding of the Universal Service Obligation

If OFCOM decides that compliance with the universal service obligations imposes a financial burden on a designated universal service provider, and that it is unfair for that provider to bear the costs of universal service alone, it may determine that contributions are to be made by service providers to which general conditions apply. OFCOM may make regulations establishing a scheme or fund for that purpose and may appoint a person to administer the scheme or fund[78]. OFCOM has not made any such regulations[79].

4.3 Access-related Conditions

OFCOM may set access-related conditions in relation to network access and service interoperability for the securing of[80]:

- efficiency on the part of communications providers and persons making associated facilities available;

- sustainable competition; and

- the greatest possible benefit for end-users.

The conditions that may be set by virtue of this authority may include:

- conditions imposing network access obligations on any person controlling access to such networks, and requiring interconnection of networks, in order to secure end-to-end connectivity for end-users[81]; and

- conditions imposing obligations on persons providing facilities for the use of application programme interfaces or electronic programme guides that OFCOM may determine are necessary to secure that persons are able to have access to such programme services provided in digital form as OFCOM may determine[82].

Sharing of Apparatus and Costs

OFCOM may set conditions for securing that persons to whom the electronic communications code applies participate in arrangements for the sharing of

---

78 Communications Act 2003, sections 70 and 71.
79 In OFCOM, *Review of the Universal Service Obligation,* 14 March 2006, OFCOM concluded that there is no unfair burden imposed on BT by the existing subsidy regime: see p. 42.
80 Communications Act 2003, section 73(2).
81 Communications Act 2003, section 74(1). These conditions may be imposed on 'any person whatever:' ibid., section 46(6). See § 4.3.1.
82 Communications Act 2003, section 74(2).
    An electronic programme guide is a list of scheduled television and radio programs typically displayed on a viewer's television screen with functions allowing the viewer to navigate and select content by time, title, channel, etc. by use of a remote control or keyboard or other device.

apparatus and apportionment of costs of shared apparatus where there are no viable alternative arrangements that may be made[83].

Technical and Operational Matters Related to SMP Conditions

OFCOM may set conditions of a technical or operational nature to secure compliance with any SMP conditions requiring an SMP operator to provide network access, and use of relevant networks and relevant facilities[84].

Conditional Access Systems

OFCOM is under a duty to apply access-related conditions to persons who provide a conditional access system[85] in relation to a protected programme service[86].

4.3.1 Access-related Conditions Imposed on BT and SSSL

Access-related conditions have been imposed on BT and Sky Subscribers Services Limited ('SSSL'). In the case of BT, OFCOM has imposed the following conditions intended to ensure end-to-end connectivity[87]:

- to purchase wholesale narrowband (fixed and mobile voice and narrowband data) call termination services from any PECN that reasonably requests in writing that BT purchases such services;

- to ensure that the purchase of the wholesale narrowband (fixed and mobile voice and narrowband data) call termination services shall occur as soon as reasonably practicable and shall be on reasonable terms and conditions (including charges), and on such terms and conditions (including charges) as OFCOM may from time to time direct:

- to ensure that after purchasing wholesale narrowband (fixed and mobile voice and narrowband data) call termination services, BT will not be able to unreasonably change, withdraw or restrict access to an applicable Normal Telephone Number; and

---

83 Communications Act 2003, section 73(3).
    The electronic communications code is set out in Schedule 2 to the Telecommunications Act 1984 as amended by Schedule 3 to the Communications Act 2003. The code is designed to facilitate the installation and maintenance of electronic communications networks. It confers rights on providers of such networks, and on providers of systems of conduits which are made available for use by providers of electronic communications networks for the purposes of the provision of those networks, to install and maintain apparatus in, over and under land and results in considerably simplified planning procedures. See also The Electronic Communications Code (Conditions and Restrictions) Regulations 2003, SI 2003 No 2553.
84 Concerning OFCOM's jurisdiction in respect of 'technical matters', see also section 4.1.2 above.
85 A 'conditional access system' is a system by which the ability to view electronic transmissions such as television programmes is limited, by means of encryption, to subscribed clients.
86 Communications Act 2003, section 75(2).
87 'End-to-end connectivity' is the process of enabling retail customers to make calls to other customers or services on the same network or other providers' networks.

- to comply with any direction OFCOM may make from time to time under this condition[88].

Access-related conditions have been imposed on SSSL in relation to the provision of conditional access services[89]. These conditions include:

- a requirement to provide conditional access services upon request, on fair and reasonable terms;

- where a broadcaster in receipt of conditional access services from SSSL also provides programme services to providers of other electronic communications networks (eg cable operators), a requirement to cooperate with providers of such other electronic communications networks so that such providers are able to transcontrol (the process of changing a conditional access system) and re-transmit the programme services;

- an obligation to keep separate financial accounts regarding activities as provider of conditional access services;

- where conditional access products and systems are the subject of intellectual property rights, a requirement to make such products and systems available upon reasonable terms and at reasonable charges (eg the licensing of technology to manufacturers of digital decoders);

- a requirement not to discriminate unduly;

- a requirement to publish charges.

4.4 Privileged Supplier Conditions

OFCOM is under a duty to ensure that 'privileged suppliers' of electronic communications comply with strict accounting rules and may impose conditions accordingly[90]. A privileged supplier is defined as a public communications provider who enjoys special or exclusive rights in relation to the provision of non-communications services and is not such a provider only in respect of associated facilities[91]. An example would be an entity which provides telecommunications

---

88 OFCOM, *End-to-end connectivity*, 13 September 2006, paragraph 4.2. This determination followed upon the decision of the Competition Appeal Tribunal in *Hutchison 3G (UK) Limited v. Office of Communications* (case no. 1047/3/3/04). The CAT referred back to OFCOM for reconsideration the question of whether H3G had significant market power in the market for mobile wholesale voice call termination on its network taking into account the extent to which countervailing buyer power exists in BT and any other matters relevant at the time of OFCOM's reconsideration. As a consequence of the CAT decision, OFCOM revisited its earlier determinations concerning the measures required to ensure end-to-end connectivity: see OFCOM, *End-to-end connectivity*, 13 September 2006, paragraph 1.2.

89 OFTEL, *The regulation of conditional access*, 24 July 2004, Annex A. Currently, only SSSL is subject to access-related conditions relating to conditional access services. However, OFCOM is currently proposing to apply identical conditions to Top Up TV Limited in respect of conditional access services provided via its DTT platform. A public consultation commenced in early 2007.

See also: OFCOM, *Provision of Technical Platform Services: A consultation on proposed guidance as to how Ofcom may interpret the meaning of 'fair, reasonable and non-discriminatory' and other regulatory conditions when assessing charges and terms offered by regulated providers of Technical Platform Services*, 2 November 2005.

90 Communications Act 2003, section 77(1) and (3).

91 Communications Act 2003, section 77(2).

services and also has an exclusive right to provide postal services. The conditions imposed in such a case can include accounting separation.

Privileged supplier conditions must be set where a public communications provider enjoys special or exclusive rights in relation to the provision of services in other sectors unless the public communications providers has an annual turnover of less than €50 million, in which case setting conditions is discretionary[92]. No UK operator has been granted any special or exclusive rights and OFCOM has not imposed any privileged supplier conditions.

4.5 Procedures for Setting, Modifying and Revoking Conditions

The procedures governing the setting, modifying and revoking of conditions are described in § 3.3.

In the case of a notification with respect to an SMP condition, some special notice requirements apply. Before OFCOM may define a market for the purposes of making a market power determination, or make a market power determination, it must publish a notification of what it intends to do. The notification must[93]:

- state that OFCOM are proposing to identify that market or to make that market power determination;

- set out the effect of the proposal;

- give their reasons for making the proposal; and

- specify the period within which representations may be made to OFCOM about their proposal.

The period for making representations must not be less than one month[94].

The Communications Act 2003 establishes requirements for the delivery of copies of notifications to the Secretary of State, the European Commission and NRAs in certain circumstances[95].

---

92 Communications Act 2003, section 77(4).
93 Communications Act 2003, section 80(3). Section 79(4)–(6) are also relevant.
94 Communications Act 2003, section 80(4).
95 Communications Act 2003, section 81.

Chapter 5

'The Undertakings'

George Ritchie

5.1 BT

5.1.2 Introduction

The development of electronic communications services in the UK over the last 30 years has been influenced by a number of significant milestones. These include privatisation and the introduction of the 1984 licensing regime, the ending of the BT/Mercury duopoly in 1991 and the advent of service provider competition in the late 1990s. On the European stage, 2003 saw full liberalisation and, in the UK, the ending of the formal licensing regime. The undertakings given by BT in 2005, under the Enterprise Act 2003 ('the Undertakings'), were undoubtedly another such milestone.

The UK has long been regarded as a European pioneer of electronic communications regulation. When the Undertakings were signed, many overseas commentators saw them as yet another of those quirky British-only developments, given under UK legislation, that would go no further—they were viewed with little more than mild amusement.

Yet within four years their impact has spread significantly and they are now seen as visionary. The central concepts of organisational separation of business units which provide bottleneck services, and of provision of such services on an 'identical' basis to that provider's downstream business and to third parties, are now recognised as being an opportunity for competitive and consumer advances rather than as a threat. They are now being taken up both in Europe and beyond.

- In Italy, Telecom Italia has established its 'Open Access' Division. In December 2008 AGCom approved 14 'Groups of Undertakings' in relation to effective separation of the access network activities from the rest of the company as well as equivalence of treatment in the provision of wholesale access services to OLOs and its own commercial divisions[1].

- In Sweden, TeliaSonera formed its network division, 'Skanova Access', in January 2008 as Sweden's telecoms regulator PTS was granted legal powers to impose functional separation. PTS's Strategic Agenda 2009[2] calls for the use of regulatory

---

1   AGCom Decision 718/08/CONS—15 December 2008. http://www.agcom.it/default.aspx?message
    =viewdocument&DocID=2688.
2   http://www.pts.se/pts/Templates/Page.aspx?id=37056&epslanguage=EN-GB.

tools to make TeliaSonera 'treat all market players that use the company's access network in the same way it treats its own end-user organisation'.

■ In New Zealand, in March 2008, Telecom New Zealand implemented a government backed plan to split into three divisions: retail, wholesale and network operations, and gave undertakings to deliver services on an equivalence basis[3].

■ In Brazil, the National Telecommunications Agency ANATEL is preparing a General Competition Goals Plan (GCGP), with the objective of establishing measures that ensure appropriate levels of competition and functional separation between networks and services.

Most notably, however, the European Commission's current proposals for reform of the 2002 Regulatory Framework include a proposed amendment which would authorise a National Regulatory Authority (NRA), in situations where other SMP remedies have failed and would persistently fail, to require a vertically integrated undertaking to place its wholesale activities in relation to the provision of access products into an independently operating business entity and to require that entity to supply access products to all on the same terms.

Practical Issue: National Regulatory Authority (NRA)

The EC proposal empowers an NRA to impose a measure which includes all the elements which are found in the arrangements briefly described above. They are:

Specification of the precise nature and level of separation, including the legal status of the separate business entity

Identification of the assets and the services to be supplied by that entity

Governance arrangements to ensure the independence of the staff employed and their incentive arrangements

Rules for ensuring compliance

Rules for ensuring transparency of operational procedures

A monitoring programme including publication of an annual report.

The existing schemes have, however, been set up largely as part of a perceived 'win-win' arrangement involving 'a new regulatory deal'; typically a refocusing of regulation upstream, with corresponding deregulation downstream anticipated.

To deliver the required outcomes, execution to both the letter and the spirit of the commitments is required. For the entity giving the commitments, anticipation of increased shareholder value as a result of delivering on the obligations may well be a key driver. It must therefore be questionable whether similar obligations imposed on an entity as a result of SMP findings, and without the vertically integrated entity's "buy-in" would be able to deliver the same levels of success.

In this chapter we review the origins and reasons for the Undertakings and their objectives. We provide a roadmap through them and an in-depth analysis of

---

3 http://www.telecom.co.nz/binarys/telecom_separation_undertakings_25_march_2008.pdf. These bear a striking resemblance to BT's undertakings.

provisions of particular interest to electronic communications providers. We end with a review of the legal framework underpinning them, including the processes for making, varying, enforcing and revoking Undertakings and appealing Undertakings related decisions.

5.1.3 Overview of BT

In order to understand how the Undertakings apply to BT, figure 5.1 below sets out an overview of the structure of BT[4]. It includes a stylised overview of how the information sharing obligations imposed on BT apply[5].

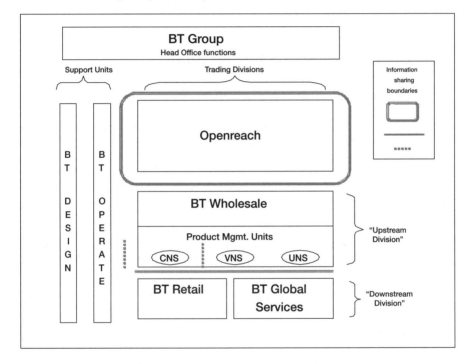

Figure 5.1—Overview of BT

BT's Annual Review 2008[6] describes the roles of the different units as follows:

- Openreach is responsible for the crucial 'first mile' connecting communications providers' customers to their local telephone exchange, giving them equal, open and economic access to the UK network.

- BT Wholesale[7] brings economies of scale to 700+ UK communications companies, through a diverse portfolio ranging from nationally available broadband, voice

---

4 As at April 2009.
5 The information sharing obligations applying to Openreach and BT Wholesale are described in sections 4.4 and 4.5 below respectively.
6 http://www.btplc.com/Sharesandperformance/Annualreportandreview/Annualreviews/AnnualReviews.htm.
7 The different product management units in BT Wholesale (CNS, VNS and UNS) and their roles are described in section 5.4.5 below.

and data connectivity services and interconnect, to bespoke, fully managed network outsourcing and value-added solutions.

- BT Retail serves consumer customers and small and medium-sized enterprises in the UK, providing a range of innovative products and services. It also comprises BT Ireland and our Enterprises division.

- BT Global Services serves corporate, carrier and government organisations across the world, providing high-performance managed networked IT services, applications management, professional services and outsourcing solutions.

- BT Design is responsible for the design and deployment of the platforms, systems and processes which support our products and services, and BT Operate is responsible for their operation.

5.2 The History of the Undertakings

5.2.1 OFCOM's Strategic Review

Within months of replacing Oftel, OFCOM heralded its arrival by announcing its intention to undertake a 'Strategic Review of Telecommunications'. It wished to move from what was perceived by some to be a sticking plaster modus operandi of multiple interventions in a range of regulatory issues to a more strategic plane which OFCOM believed would better enable it to deliver its objectives under the Communications Act 2003.

In March 2004 OFCOM published the first consultation document of its Strategic Review[8]. This was positioned as a fact finding inquiry into the nature of the market, but it contained five questions that gave a clear view of OFCOM's intent. These were:

- **Question 1:** In relation to the interests of citizen-consumers, what are the key attributes of a well-functioning telecoms market?

- **Question 2:** Where can effective and sustainable competition be achieved in the UK telecoms market?

- **Question 3:** Is there scope for a significant reduction in regulation, or is the market power of incumbents too entrenched?

- **Question 4:** How can OFCOM incentivise efficient and timely investment in next generation networks?

- **Question 5:** At varying times since 1984, the case has been made for structural or operational separation of BT, or the delivery of full functional equivalence. Are these still relevant questions?

It was clear that structural separation was on the agenda as a threat, and deregulation as an incentive.

OFCOM published its findings in November 2004 in its second consultation document[9]. It concluded that whilst competition at the retail level was bringing

---

8   http://www.ofcom.org.uk/static/telecoms_review/condoc_phase1.htm.
9   http://www.ofcom.org.uk/static/telecoms_review/condoc_phase2.htm.

significant benefits to consumers, especially business customers, the picture was very different at the wholesale level. OFCOM's view was that the present market conditions were unsustainable but, taking note of industry concerns about the likely period of uncertainty and the potential costs of pursuing a break up of BT, hinted strongly that a solution short of a referral to the Competition Commission on structural separation might be preferable, provided that the right refocusing of regulation could be achieved on a voluntary basis.

OFCOM's prime focus was on the markets that it considered to be economic bottlenecks, in particular the supply of access and, to a degree, backhaul services which were considered likely to endure. Its view was that if regulation was properly focused here, it might be possible to withdraw from downstream regulation where competition was more vibrant.

> *'In local access and other wholesale access products, efficient and sustainable competition is likely to require some continuing regulation to secure genuine equality of access, right through from product design to customer handover. Such regulation needs to be focused on a more limited range of wholesale products than to date—where there are real bottlenecks that are likely to endure. However, where it is focused, it also needs to be more intensive than hitherto. Such an approach, of much more tightly focused but intensive intervention to guarantee genuine equality of access through key bottlenecks, also creates real scope for a significant withdrawal from sector-specific regulation.'*[10]

OFCOM also set down the principles to which it would operate in the future:

> *'The principles we propose to guide our actions are:*
> 1. *promote competition at the deepest levels of infrastructure where it will be effective and sustainable;*
> 2. *focus regulation to deliver equality of access beyond those levels;*
> 3. *as soon as competitive conditions allow, withdraw from regulation at other levels;*
> 4. *promote a favourable climate for efficient and timely investment and stimulate innovation, in particular by ensuring a consistent and transparent regulatory approach;*
> 5. *accommodate varying regulatory solutions for different products and, where appropriate, different geographies;*
> 6. *create scope for market entry that could, over time, remove economic bottlenecks; and*
> 7. *in the wider communications value chain, unless there are enduring economic bottlenecks, adopt light-touch economic regulation.'*[11]

5.2.2 Delivery of the Undertakings

BT's response to the second consultation document offered organisational and behavioural remedies including the principle of 'equivalence of access' to bottleneck services. A period of intense negotiations followed, in the course of which BT agreed

---

10 Foreword, page 5, Strategic Review of Telecommunications Phase 2 consultation document: http://www.ofcom.org.uk/consult/condocs/telecoms_p2/tsrphase2/maincondoc.pdf.
11 Paragraph 1.25 Strategic Review of Telecommunications Phase 2 consultation document.

to embody its proposals in voluntary Undertakings to OFCOM in lieu of a reference to the Competition Commission under the Enterprise Act 2002.

On 30 June 2005, OFCOM issued a consultation document under Section 155 Enterprise Act 2002[12]. This is useful in that it identifies the market in which OFCOM had found market failure: At Section 3.5, OFCOM stated:

> *'... OFCOM considers that the set of markets in which competition is adversely affected by the combination of features it has identified consists of the upstream markets for the provision of access and backhaul network services and all related downstream markets, including retail markets, for which the former services are a critical input.'*

It also sets out OFCOM's rationale for the acceptance of the various commitments. Following consultation and minor amendments to the draft text, the Undertakings came into effect on 22 September 2005.

5.2.3   Evolution of the Undertakings

BT and OFCOM both contemplated, when the Undertakings were given, that changes would be needed in the light of both technological and market developments. The Undertakings were also intended to be, to some extent, high level principles rather than a re-write of old style licence conditions. As a result, there are a considerable number of commitments that apply 'unless OFCOM consents otherwise' or which may be varied by agreement between BT and OFCOM. These provisions have proved apt and necessary as both the creation of Openreach (and with it the introduction of 'Equivalence of Input'[13] or EOI) and the obligations relating to systems separation have resulted in significant change programmes for BT. The scale of these change programmes, coupled with technological and market place developments (some of them direct consequences of the Undertakings themselves) have meant that the Undertakings have had to evolve and will undoubtedly continue to do so.

OFCOM's website[14] contains a copy of the Undertakings, a full list of exemptions and variations[15] and other key documents. In effect, it tracks the evolutionary journey of the Undertakings.

A 'Variation' is a formal change to the Undertakings which is permanent. In the three years from signature to the end of 2008 there had been 18 Variations. Some have been to refocus commitments in the light of subsequent knowledge of the parties—for example Variation 16[16] which, inter alia, redefined the systems separation requirements in the light of representations from BT as to the complexity of this. Others such as Variation 17[17]—removal of the EOI obligations in broadband Market 3—have been to withdraw obligations where they are no longer needed because the markets are now competitive and where the Undertakings have fulfilled their purpose.

---

12   http://www.ofcom.org.uk/consult/condocs/sec155/.
13   The principle of EOI is explained in section 5.3.2. (i) below.
14   http://www.ofcom.org.uk/telecoms/btundertakings/.
15   http://www.ofcom.org.uk/telecoms/btundertakings/exemptionsandvariations/.
16   http://www.ofcom.org.uk/telecoms/btundertakings/exemptionsandvariations/statement071008. pdf.
17   http://www.ofcom.org.uk/telecoms/btundertakings/exemptionsandvariations/wavestream1208.pdf.

An 'Exemption', in contrast, leaves the underlying Undertakings commitment unaltered, but provides (usually temporary) relief from the requirement to comply with a particular commitment. For example, in June 2006, OFCOM agreed to an exemption lasting three months in relation to the operation of the BT Wholesale broadband line checker[18].

There have also been other changes, such as the movement of product management of various BT Wholesale broadband products, which was effected in July 2008 through the change mechanism expressly built into Section 6.4 involving an exchange of letters[19].

5.2.4 Staying up to Date with the Undertakings

In December 2008, OFCOM published a consolidated version of the Undertakings and agreed with BT that in future it would aim to publish a revised consolidated version at the same time as agreeing future variations. However, to be sure that you have the correct text, it is recommended that you should:

- Check the most up to date consolidated version of the Undertakings published by OFCOM[20].

- Check OFCOM's Undertakings Implementation website[21] for any subsequent Variations.

- Review OFCOM's Undertakings Implementation website for any exemptions or other agreements that may 'over-ride' the substantive Undertakings commitment in question.

5.3 An Undertakings Overview—the Concepts, Themes and Route Map

The Undertakings need to be understood as a tool to deliver particular ends (rather than as an end in their own right). In particular they are intended to remedy the concerns identified by OFCOM in the Strategic Review, to help the achievement of the principles and objectives described above and to do this by delivering organisational and behavioural change.

For this reason, rather than embarking on an end to end read of the Undertakings, or 'diving straight in' to specific sections of them, it is useful to have an overview of the general shape of them, and what in particular the different sections are intended to deliver. With an understanding of the framework of the Undertakings, navigation through the detail of over 250 individual commitments becomes far easier.

The following paragraphs do two things to give the reader the ability to find what they are looking for and to understand the details of the rules. First, they give a high

---

18   http://www.ofcom.org.uk/telecoms/btundertakings/exemptionsandvariations/ofcom204.pdf.

19   http://www.ofcom.org.uk/telecoms/btundertakings/exemptionsandvariations/070808_
     UndertakingsIPStream.pdf.

20   http://www.ofcom.org.uk/telecoms/btundertakings/ The consolidated version should show up to
     which number variation it includes.

21   http://www.ofcom.org.uk/telecoms/btundertakings/exemptionsandvariations/.

level overview of the concepts which underpin the organisational and behavioural change commitments. Secondly, they set out a section by section route map through the Undertakings.

5.3.1 Organisational Change

(i) The Creation of Openreach

The creation of Openreach as the supplier of BT's Network Access and Backhaul services is the most evident organisational change brought about by the Undertakings. Openreach's establishment as, in effect, 'a company within a company' is the key enabler of trading between Openreach and other parts of BT on the same 'arms length' basis as occurs with other communications providers pursuant to the Equivalence of Inputs (EOI) commitments (see Part 5.3.2 below). Section 5 of the Undertakings deals with the creation of Openreach, the activities which must be undertaken only by people located in Openreach, and other structural requirements that dictate how Openreach must behave and trade.

(ii) Changes Within BT Wholesale

Having regard to the nature and extent of competition in the wholesale space downstream of the access network, the Undertakings also required a degree of organisational restructuring in BT Wholesale; in particular of the product management units to separate the product management of upstream and downstream wholesale voice and data products. Section 6 of the Undertakings deals with the BT Wholesale obligations.

(iii) Account Management and Sales Functions

In short, the Undertakings give communication providers dealing with BT the right to choose where in BT they wish to be account managed from and impose divides between different sales functions. Separation of account management and sales functions is dealt with in Sections 5, 6 and 8.

(iv) Systems Separation

The Undertakings mandate separation of BT systems, but this is very much 'work in progress' and will be until at least June 2010. The main focus here is the intention that (unless alternative solutions are agreed with OFCOM) by June 2010 Openreach will physically separate its Operational Support Systems (OSS) and will implement a lesser degree of systems separation of its Management Information Systems (MIS). See Sections 5, 8.5 and Annexes 5 and 6.

Systems separation has been one of the costliest aspects of the Undertakings for BT and the subject of a number of Variations. The initial text of the Undertakings called for logical separation of Openreach MIS and OSS at a relatively early date with physical separation by 2010. However BT and OFCOM views of what was meant by logical and physical separation were not thereafter fully aligned and so the Undertakings have evolved to include a set of requirements based on differing degrees of separation and milestones along the way to full physical separation.

In December 2008, recognising the cost of implementation of the systems separation obligations, the changed economic environment, and the evolving needs of Communications Providers, OFCOM announced that it was launching a project:

> *'to consider the scope for Openreach to refocus its systems-related resources in a way that accommodates new, high priority demands placed on Openreach's systems resources whilst safeguarding the outcomes envisaged by the Undertakings'*[22].

Further changes to the systems separation requirements may therefore evolve during 2009.

5.3.2 Behavioural Change

(i) Equivalence of Inputs ('EOI')

EOI is best described as 'the same means the same'. A key concern identified by respondents to OFCOM's Strategic Review was that on a range of issues, the way in which BT dealt with downstream customers internally and externally differed. None of these were in their own right so material as to amount to regulatory undue discrimination which breached Significant Market Power ('SMP') conditions[23], but taken together they had a competitive impact.

EOI is defined in Section 2 of the Undertakings and is applied to a range of important products supplied by BT as listed in Section 3 of the Undertakings. The main focus for EOI is Openreach and Section 3 lists the main products supplied by Openreach and consumed by other Communications Providers, in particular those who are investing in local loop unbundling. Section 3 does however also extend to cover certain BT Wholesale broadband access products of particular significance to other Communications Providers.

Whilst it was accepted when the Undertakings were signed (having regard both to the greater competition for the provision of core network services and to the cost that would be involved) that it would be inappropriate to require BT's core network to be re-engineered so that it could be provided on an EOI basis, it was recognised that BT was planning to build a Next Generation Network (NGN) and that there were risks that this could be constructed in ways detrimental to the interests of other Communications Providers. Section 11 accordingly sets out the principles that BT must build into its NGN, including the principle that services provided by means of the NGN should be provided on an EOI basis if BT has Significant Market Power for the supply of those services.

(ii) Information Sharing

The concern was raised in the Strategic Review that information flows within BT gave it a competitive advantage and in particular that those parts of BT that

---

22 http://www.ofcom.org.uk/telecoms/btundertakings/otherdocs/keyIT.pdf.
23 See chapter 4.

consumed upstream product inputs also used by other Communications Providers could inappropriately influence the commercial policy of those parts of BT that supplied those upstream products.

The Undertakings solution is to define 'Commercial Information' ('CI') —information about BT's wholesale SMP products, 'Customer Confidential Information' ('CCI') — information shared with BT by its customers, and 'Commercial Policy' ('CP') —BT's plans for its wholesale SMP products[24]. Section 5 of the Undertakings imposes the strictest rules for the sharing of Openreach CI and CCI, and on the influencing of Openreach CP, whilst Section 6 contains the slightly less onerous rules regarding BT Wholesale CI, CCI and CP. Section 8 contains provisions in relation to the treatment of CI and CCI by sales forces of different parts of BT. The approach of all the rules is, in effect, to limit information flows between different BT lines of business and to prohibit the influencing of CP other than through mechanisms available to all Communications Providers.

Given that BT remains a vertically integrated company, there remains a need for information flows to be able to transcend organisational boundaries for a range of managerial and governance purposes. Annex 2 identifies functions, and therefore the individuals undertaking those functions, who have the right to see across the boundaries created by the organisational separations and who are therefore not bound by the organisational information sharing rules described above. There are two parts to Annex 2. Annex 2 Part A people can see CI and CCI and may seek to influence CP. Annex 2 Part B people may see CI and CCI but may not seek to use their position to influence CP. The Board of BT plc, Legal and Regulatory functions, Group Risk and Insurance, and Head of Ethics are examples of Annex 2, Part A functions. Annex 2 Part B functions include Compliance and Internal Audit, Group Technology, BT Property and Press, Communications, and Investor Relations

Finally, Section 11.20 provides that the information sharing rules shall not impede the flow of information required to enable BT to design, build and implement BT's Next Generation Network or the decision making process relating thereto.

(iii)   Compliance

The Undertakings contain a mesh of commitments designed to drive the required behavioural changes and ensure that BT is committed to compliance. They include commitments to make goodwill payments if certain commitment milestones were not delivered (Section 3.2), the requirement for Codes of Practice for all BT people (Section 9) and the establishment of the Equality of Access Board as an entity responsible for oversight of delivery and delivery of transparency externally of BT's achievements (and failures) (Section 10).

5.3.3   An Undertakings Route Map

Those drafting the Undertakings sought to adopt a logical order for ease of reference and navigation. The summary below gives an overview of the different sections and is intended to provide an overview of where to find particular topics. In Part 5.4 below the key provisions are considered in greater detail.

---

24   Fuller descriptions of these terms are set out at Part 5.4.1 below.

Section 1: Scope. This defines the scope of application of the Undertakings.

Section 2: Definitions and Interpretation. This contains over 90 defined terms used in the Undertakings and other matters of interpretation. Some of the definitions are lengthy and contain 'the meat' of what a commitment applies to.

Section 3: Provision of Equivalent Products and Services. This imposes the EOI obligations for products existing and anticipated when the Undertakings were given and rules for the application of EOI to new products. It links to Annex 1 which contains the timetables for the introduction of EOI.

Section 4: Transparency. This contains specific rules relating to Partial Private Circuits, Carrier Pre-selection, the BT product DataStream and how new TILLAP and TILLBP products[25] should be treated.

Section 5: Access Services. This contains the Openreach specific provisions— Openreach being BT's trading name for the entity known in the Undertakings as BT's Access Services ('AS') division. It deals (inter alia) with the establishment of Openreach, which products it will provide, what assets it should control and operate, the principles for the provision of backhaul services, the composition of Openreach—including what people should be in it—and the principles for their remuneration. It also covers Openreach's duties, governance, and financial reporting, its information sharing rules (subject to the provisions of Annex 2), its systems separation requirements, its requirement to offer services to the rest of BT on an EOI basis, its sales activities and its brand.

Section 6: Management and Structure of BT Wholesale. This sets out the requirements in relation to the product management of BT Wholesale's SMP and Significant products[26], remuneration principles for certain BT Wholesale people and BT Wholesale's information sharing rules (subject to the provisions of Annex 2). It also introduces obligations in relation to the product management of certain leased lines products in accordance with the obligations set out in Annex 3.

Section 7: Equipment Location. This deals with the provision of space in BT's exchanges and needs to be read in conjunction with the definition of Equipment in Section 2 and the provisions of Annex 4.

Section 8: Separation of Upstream and Downstream Divisions. This tidies up some residual obligations in relation to organisational separation, most notably in relation to sales functions and information sharing. It overlaps with Section 6.

Section 9: Code of Practice. This introduced the compliance obligation on BT to produce codes of practice to help BT people understand what was required of them and other compliance training.

Section 10: The Establishment of an Equality of Access Board. This establishes the EAB, including its composition, role and duties. It provides that the essence of the EAB is that it is a robust, independent body to validate and report on delivery of and adherence to the Undertakings.

---

25 Traditional Interface Leased Line Access Product and Traditional Interface Leased Line Backhaul Product—new products to be developed by Openreach as defined in Section 2 of the Undertakings.
26 The product management units and the 'Significant products' are explained at Part 5.4.5 below.

'The Undertakings'

Section 11: Next Generation Networks (NGN). This contains the principles to apply to the design and operation of BT's NGN as and when BT builds it. It contains provisions on network design, the requirement for services provided over the NGN to be provided on an EOI basis where there is SMP and prohibiting the foreclosure of network access. It includes provisions for consultation with industry groups, in relation to an operational dispute adjudicator, and sets out the principles that BT will consider for paying compensation in particular circumstances.

Section 12: Contract Management. This was a time limited (six months) provision in relation to setting up pan-industry contract management mechanisms.

Section 13: Northern Ireland. This provides that most provisions of the Undertakings do not apply to BT in Northern Ireland where it remains organisationally unaltered. This 'carve-out' was agreed because the organisation in Northern Ireland was too small to warrant the costs associated with effecting the types of separation envisaged by the Undertakings.

Section 14: Information Requests and Co-operation. This entitles OFCOM to make information requests to BT which, if not complied with, would be a breach of the Undertakings.

Section 15: Directions. This contains what may be described as 'half-way house' enforcement provisions enabling OFCOM to serve BT with an enforcement direction (subject to procedural safeguards) if it is satisfied that BT has breached the Undertakings[27].

Section 16: Breach of the Undertakings. This deals with requirements for BT to obtain consent or agreement, stating that failure to do so where required will be a breach, but that such consent or agreement will not be unreasonably withheld or delayed.

Section 17: Compliance with Other Legal Requirements. This provides an important safeguard for BT in that it establishes that where other legislation requires BT and its people to do something, the Undertakings do not override that obligation or prevent its delivery.

Section 18: Variation of the Undertakings. This confirms that the Undertakings may be varied by agreement between BT and OFCOM but is silent on process.

Section 19: Expiry and Termination. This provides the framework for the falling away of the Undertakings, either through markets becoming competitive or for other reasons[28].

Section 20: General. This sweeps up a number of procedural 'loose ends'. It also contains specific provisions in relation to 'Exceptional Incidents' (added since the original signature of the Undertakings) to provide for flexibility in times of local or national disaster.

Annex 1: Equivalence of Inputs Timetable. This sets out the timetable for the introduction of EOI for the products listed in Section 3 of the Undertakings. Typically it includes

---

27 The provisions of Section 15 are considered under Part 5.5 below (5.5.3 Enforcement of the Undertakings).
28 The provisions of Section 19 are considered under Part 5 below (5.5.4 Revocation of the Undertakings).

the Ready for Service date (the date when the product must first offered to certain customers on an EOI basis) and the Installed Base Migration Complete date (the date by which provision of that service on an EOI basis to all customers must be achieved).

Annex 2. This sets out the individuals and groups of people who may see CI and CCI (Part A and Part B people) and those who may also influence CP (Part A people only) when the information sharing rules would otherwise prevent them from doing so.

Annex 3: Leased Lines. This contains specific provisions in relation to the provision by BT Wholesale of communications provider variants of certain specified retail leased lines.

Annex 4: Equipment. This lists the types of equipment to which Section 7 of the Undertakings applies.

Annex 5: List of the Thirteen MIS Systems. This lists the specific BT Management Information Systems to which the systems separation requirements apply.

Annex 6: List of the Operational Support Systems. This lists the Operational Support Systems that contain a view of BT's network inventory and which it is not appropriate to separate but which are to become subject to User Access Controls pursuant to Section 5.44.6.

5.4 Key Sections of the Undertakings—In Depth

5.4.1 Section 2—Definitions

This section of the Undertakings is often overlooked, but can be a vital aid to interpretation. Many of the definitions are relatively self-evident, or refer out to established definitions already set out elsewhere, for example, in the Communications Act 2003 or international standards. Definitions of particular significance, grouped by reference to specific Undertakings themes, include the following.

(i) Product and Network Related Definitions

The Undertakings define three types of nodes. A '**Local Access Node**' means a node which supports the provision of services to End-Users. A '**Core Node**' means a node whose primary function is not to support the provision of access services to End-Users, but to switch or route traffic between other nodes in a network. An '**MSAN**' means a multi-service access node being a Local Access Node in BT's NGN which is capable of supporting the provision of multiple services to end-users whether over copper or fibre.

'**Access Network**' means the Electronic Communications Network which runs from a Local Access Node to a network termination points on an End-User's premises and which supports the provision of copper or fibre-based access services.

'**Backhaul Product**' means a Network Access service which runs from a BT Local Access Node to another BT Local Access Node or a BT Core Node or another Communications Provider's point of handover—provided that its straight line

105

distance is no more than the greater of 15km or the distance from BT's Local Access Node to the nearest Core Node.

'BT's Backhaul Network' means BT's Electronic Communications Network from BT's Local Access Nodes to another BT Local Access Node or a BT Core Node or another Communications Provider's point of handover.

'NGN' means Next Generation Network, a packet-based Electronic Communications Network which is able to provide Electronic Communication Services and to make use of multiple broadband and quality of service-enabled transport technologies, and in which service-related functions are independent of underlying transport-related technologies.

Various BT products are also defined including **Backhaul Extension Service, DataStream, Featureline, IPStream, IPstream Connect, Wholesale Calls, Wholesale Extension Service**, and **Wholesale Line Rental**.

(ii)   Equivalence of Inputs Related Definitions

The definition of **'Equivalence of Inputs'** is central to the Undertakings. The definition is lengthy, but has at its heart a requirement for BT to provide in respect of a particular product or service, the same product or service to all Communications Providers (including BT) on the same time scales, terms and conditions (including price and service levels) by means of the same systems and processes. It includes the provision to all Communications Providers (including BT) of the same Commercial Information about such products, services, systems and processes. In particular, it includes the use by BT of such systems and processes in the same way as other Communications Providers and with the same degree of reliability and performance as experienced by other Communications Providers.

'The same' means exactly the same save only for very limited exceptions listed in the definition. These make allowance for the fact that BT is a single entity and therefore cannot, for example, subject other parts of itself to credit vetting or legally binding contracts.

'RFS Date' and **'IBMC'** are definitions which relate to the transitional process whereby a particular product moves from non-EOI delivery to EOI delivery. The Ready For Service (RFS) date is the date on which an EOI service must be available to both BT and other Communications Providers and be in use by BT for New End-Users for the retail service which consumes the EOI product. The Installed Base Migration Complete (IBMC) date is the date by which all of the existing installed customer base must have been migrated over so as to consume the EOI service.

'End-User' is defined by reference to the Communications Act 2003 and, together with the description of what is a New End-User as set out in the RFS date definition, is of relevance in determining which customers must be supplied on an EOI basis between the RFS date and the IBMC date.

(iii)   Information Sharing Related Definitions

'Commercial Information' ('CI') means information of a commercially confidential nature relating to SMP products, EOI products and Significant products under Section 6, and which relates to product development, pricing, marketing strategy and intelligence, product launch dates, cost, projected sales volumes or network coverage and capabilities.

'**Commercial Policy**' means policies and plans relating to the same product set as above, and which relates to product development, pricing, marketing strategy and intelligence, product launch dates, cost, payment terms, product specific forecasting or network coverage and capabilities, but excludes anything agreed with OFCOM or commercial policy of general application across BT which it is appropriate to set centrally.

'**Customer Confidential Information**' ('CCI') means information in any form which if written or electronic is clearly designated by the Communications Provider as commercially confidential and if orally disclosed is identified as such or is by it nature commercially confidential. It excludes information in the public domain (unless there by reason of a confidentiality breach), any information which was previously known by BT at its time of receipt, or which was independently generated or discovered by BT or was subsequently received from a third party without any restriction on disclosure.

(iv) Systems Separation Related Definitions

Systems separation has probably spawned more definitions—and more Variations—than any other single Undertakings subject.

The types of systems to be separated are defined. They are:

- '**Management Information Systems**' (MIS) which are the systems that hold CI and/or CCI and which are used by BT to help plan and direct business and organisational operations, decision making and competitive strategies; and

- '**Operational Support Systems**' (OSS), meaning those support systems carrying out the functions and processes which help to run a network and business and including (but not limited to) pre-ordering, taking a customer's order, configuring network components, creating a bill and managing faults.

The types of separation are also defined. The definitions include:

- '**User Access Controls**' which involves using user profiles and authorisation control mechanisms so that only certain users can see particular types of data;

- '**Level 1 Systems Separation**' which involves application of access rights and controls to restrict access to information on a system and, in the case of OSS, functionality; and

- '**Level 2 Systems Separation**' which is both separation of data held in the system and separate instances of the application software.

The systems separation provisions involve a roadmap to full delivery with milestones along the way. That roadmap is loaded with definitions of the types of records to be moved including **Customer Service Record, Customer Side Record**, and **Supply Side Records** and process terms such as **Measured Products** and **Ready to Mass Migrate**. The detail is likely to be of interest to few outside OFCOM and BT.

5.4.2 Section 3—Provision of Equivalent Products and Services

Section 3.1.1 of the Undertakings lists the specific products to which the EOI obligation applies. Those products are:

■ Wholesale Analogue Line Rental;

■ Wholesale ISDN2 Line Rental;

■ Wholesale ISDN30 Line Rental;

■ Wholesale Extension Service;

■ Shared Metallic Path Facility (SMPF);

■ Metallic Path Facility (MPF);

■ IPStream;

■ IPstream Connect[29]; and

■ Backhaul Extension Service.

BT's obligation to supply those products on an EOI basis is governed by the timetable of RFS and IBMC dates set out in Annex 1.

If a product is not listed in this section, and is not supplied by Openreach to BT, there is no obligation for EOI supply. A product or service not listed in Section 3 and supplied by Openreach to BT must, however, be supplied equivalently unless expressly exempted (see Section 5.46.1).

The majority of the products listed in Section 3.1.1 are enduring economic bottlenecks. The inclusion of the non-Openreach products IPStream and IPstream Connect and Wholesale Line Rental (which are not enduring economic access or backhaul service bottlenecks) in the Section 3.1.1 list merits explanation.

The enduring economic bottleneck in the broadband value chain is an unbundled loop. This is recognised in the requirement for EOI supply of MPF and SMPF services and in the RFS dates set for IPStream to consume them. However, when the Undertakings were signed, take-up of LLU (MPF and SMPF) services was very low, consumption of IPStream was high and inclusion of IPStream was a compromise which was expedient given the then prevailing market conditions.

Similarly, inclusion of Wholesale Line Rental was a compromise reached after considerable negotiation. Whilst it was accepted that Wholesale Line Rental was not an enduring economic access or backhaul service bottleneck, OFCOM felt unable, when the Undertakings were agreed, just to rely on MPF and SMPF given that the vast majority of BT's competitors still relied heavily on it.

Section 3.1.2 specifies future products which must be provided on an EOI basis. To the extent that it lists Openreach products, arguably it is redundant given Section 5.46.1.

The remainder of Section 3 is a combination of a mop up of specific product issues which had emerged and were amenable to early resolution, the last-minute

---

29 This product was added by Variation 14.

introduction of RFS dates for analogue WLR and MPF and SMPF and penalties for missing them, and a clarification that where BT has an obligation to apply EOI to a product it includes internal and external migration processes.

The Undertakings were written in an environment where SMP applied nationally. In May 2008, at the conclusion of the Wholesale Broadband Access Market Review, OFCOM found there to be different geographic markets. BT was found to have SMP in Markets 1 and 2 but not in Market 3. Whilst OFCOM expressed a view that there was no direct linkage between Section 3.1 of the Undertakings and SMP, it did nonetheless agree to vary Section 3 of the Undertakings so as to remove the EOI obligation in relation to the supply of the wholesale broadband products IPStream and IPstream Connect in Market 3, saying that it regarded this variation as being not material on the basis that it was a consequence of the findings of competitiveness in Market 3 in the Wholesale Broadband Access Market Review[30].

At the conclusion of the Business Connectivity Market Review in December 2008, OFCOM found that in relation to the supply of AISBO[31] services above 1Gb, BT does not have market power. We can therefore anticipate the possibility of further changes to the Undertakings in 2009 to remove the EOI obligations in those areas, on the same basis.

5.4.3 Section 4—Transparency

During industry discussions on the Strategic Review, concerns were raised over the degree to which some non-bottleneck services (Partial Private Circuits, Carrier Pre-selection and DataStream) were provided without undue discrimination. The normal SMP requirement was for the publication of an internal reference offer (IRO) which would show Communications Providers how the services they were buying were used as part of the cost and product build of BT's retail services. These IRO's were mistrusted and OFCOM and BT agreed to include within the Undertakings provisions requiring improved transparency.

5.4.4 Section 5—Access Services: Openreach

This section defines Openreach, its assets and portfolio and deals with its governance, incentives and responsibilities. Sections 5.1 to 5.15 deal with the establishment of Openreach.

Section 5.3 makes it Openreach's responsibility to provide those SMP products which are predominantly provided using the Physical Layer or Transmission Layer of BT's Access Network or Backhaul Network. It specifies that the provision of products includes product management, sales, in life service management, specification of the products and the setting of prices.

By Section 5.4 Openreach inherited the EOI portfolio listed in Section 3.1.1 with the exception of IPStream and IPstream Connect which are product managed by

---

30 See Variation 17: http://www.ofcom.org.uk/telecoms/btundertakings/exemptionsandvariations/ wavestream1208.pdf.
31 Alternative Interface Symmetric Broadband Origination: A form of symmetric broadband origination service providing symmetric capacity between two sites, generally using an Ethernet IEEE 802.3 interface.

BT Wholesale. It is of note that the obligations in relation to MPF and SMPF include 'Associated Services' such as co-mingling space in exchanges and other mechanical and electrical needs. Section 5.5 required Openreach to offer a future product set (which specifically includes sub-loop unbundling). By virtue of Section 5.8, this list is not exhaustive.

Section 5.7 contains a rather obscure reference to network access and MSANs echoing a similar statement in Section 3.1.2 which reflects the uncertainty which existed when the Undertakings were signed with regard to the development of BT's NGN and MSAN functionality. It is of relevance only to those concerned with the design of BT's NGN and the services that run over it.

Section 5.11 makes it clear that Openreach can choose whether or not to develop services above and beyond those required under SMP rules, but that it must do so under a fair statement of requirements process. This is however subject to the constraints which arise as a result of the definition of the asset base of Openreach as set out in Sections 5.12 and 5.13.

Section 5.12 makes Openreach responsible for controlling and operating the assets contained within the Physical Layer of BT's Access Network and Backhaul Network, including those items needed to support these assets, such as line testing and remote diagnostics. By Section 5.13 Openreach shall not control and operate the assets contained within the transmission layer of BT's Access Network or Backhaul Network save as permitted by Section 5.51 which allows Openreach to control and operate the electronic equipment necessary to provide super-fast broadband services over FTTC (fibre to the cabinet). The phrase 'control and operate' is used to describe Openreach's stewardship of its assets. This is a surrogate for 'ownership' which recognises that the assets are BT's and that Openreach is an organisational unit rather than a discrete legal entity.

Section 5.13 sets out ways in which Openreach can ensure that the transmission assets it needs are provided and developed in a way which enables it to discharge its responsibilities. Section 5.13.4 requires Openreach to use a statement of requirements process for new product developments. This requirement needs to be read in conjunction with Section 5.43 which obliges Openreach to evaluate new product requests in a way which is substantially the same for all communications providers for EOI products and is on a not unduly discriminatory basis in the case of other SMP products.

Section 5.14 covers the people to be included in Openreach. It includes most field engineers, those employees involved in the design, planning, implementation and in life service management of Openreach products and ancillary staff. It needs to be read in conjunction with Section 5.42 which allows Openreach to draw support from elsewhere in BT provided that the information sharing rules are respected.

Section 5.15 introduces a limited degree of flexibility in the use of engineering resources (up to 250 engineers) in specific locations. This has regard to the challenges of economically serving remote geographies such as the Scottish islands and the Isles of Scilly, when it would be inefficient—and result in under utilisation—if there were two different workforces.

Sections 5.16 to 5.18 give a number of specific directions to the shaping and pricing of the Openreach backhaul portfolio. They include provisions on aggregation, including the use of daisy chains and hub and spoke topologies designed to drive network efficiency, but not where the use of such topologies would have the intent or effect of replicating a core network.

Sections 5.20 and 5.21 recognise that the obligations relating to Openreach stem from findings of Significant Market Power and therefore provide for them to fall away as SMP is lost by giving BT the choice of moving Openreach assets used to supply non-SMP services out of Openreach, and by allowing reallocation of product management outside Openreach.

Sections 5.23 to 5.29 deal with the composition and governance of Openreach as a separate business division and are designed to ensure that it will have appropriate autonomy and, as a safeguard, provide for the Equality of Access Board to have oversight of the Openreach annual operating plan.

Sections 5.30 to 5.32 cover the nature of, and timing for, Openreach accounting. The objectives here are twofold: to ensure transparency of Openreach financial information and to secure consistency with, or an explanation of divergence from, the regulatory accounts which BT publishes annually.

Sections 5.33 to 5.37 deal with the geographic separation of senior Openreach people, demarcation of individual's responsibilities and separation of their incentive remuneration (bonuses) to ensure that they do not have divided loyalties.

Sections 5.38 to 5.42 deal with what can generically be described as 'information sharing' by Openreach. The rules cover three subjects—all defined terms—Commercial Information, Commercial Policy and Customer Confidential Information[32] and reflect a series of compromises between OFCOM and BT, given that OFCOM wanted the tightest rules possible, but BT wished to ensure that corporate accountability and controls were not jeopardised.

By Section 5.38, there is a prohibition on any non-Openreach people seeking to participate in the making of Openreach Commercial Policy, or influencing or seeking to influence it, 'except through such mechanisms and processes that are also available to other Communications Providers'. There is also a prohibition on any non Openreach person having access to Openreach Commercial Information 'unless it is of the nature that would be provided to other Communications Providers in the ordinary course of business'.

Section 5.41, however, introduces Annex 2 which provides for various individuals and classes of people to be exempted from the information sharing rules. Annex 2 Part A people may both see Commercial Information and seek to influence Commercial Policy. Annex 2 Part B people may only see Commercial Information.

Section 5.39 provides that Openreach Customer Confidential Information shall not be disclosed to employees working for the other business divisions, BT Wholesale, BT Retail, or BT Global Services). In relation to the support units, it shall not be disclosed to BT Operate, or to some people in BT Design (ie those BT Design people whose incentive remuneration reflects the objectives of the other business divisions). Openreach Customer Confidential Information can therefore be disclosed to people in BT Group. As with other types of commercial information, there are exceptions

32   The definitions are detailed and are explained at Part 5.4.1 above.

where disclosure is permitted. These include disclosure to Annex 2 people, where required to enable operational delivery of Openreach products and most importantly of all, when the customer consents to the sharing of its information.

Section 5.40 is the corollary to Section 5.38.1. It prohibits Openreach (non–Annex 2) people influencing or attempting to influence the Commercial Policy of the other BT business divisions save through such mechanisms as are available to other Communications Providers and where it is required for the delivery of the Undertakings; for example for the development of Openreach products which depend on the use of network assets outside Openreach's control.

Section 5.44 deals with separation of OSS systems and Section 5.45 deals with separation of MIS. They are linked with Section 8.5.

The basic intention is that at an operational level, Openreach should have its own systems, and so the over-arching requirement in Section 5.44.1 is that BT shall ensure that its OSS designed for Openreach are designed on the principle of separation from the rest of BT and Section 5.44.2 requires physical separation by 30 June 2010. This is however caveated by the exception of the OSS listed in Annex 6 for which, by Section 5.44.6, the imposition of User Access Controls as described therein is sufficient.

Section 5.44.3 then provides a roadmap setting out the milestones along the way to full physical separation of the relevant OSS. This roadmap refers firstly to the implementation of User Access Controls for OSS supporting listed EOI products on dates between June 2007 and February 2008. These controls are now in place. It then sets out milestones for moving Customer Side Records and Supply Side Records relating to specific products to physically separate systems. These milestones are dates by which either 50 per cent and then 90 per cent of the relevant type of records for the listed products must have been migrated to physically separate systems. Sections 5.44.4 and 5.44.5 provide for the roadmap and implementation to be kept under review by OFCOM and subject to EAB scrutiny.

In relation to MIS, the requirement is slightly less onerous. Rather than requiring full physical separation, lesser degrees of separation, described as Level 1 System Separation and Level 2 System Separation suffice. BT was required to implement Level 2 System Separation of its MIS, except for those systems listed in Annex 5 within 13 months. For the Annex 5 systems, Level 1 System Separation sufficed within 13 months, but BT is required to implement Level 2 Systems Separation by June 2010, unless the systems are no longer shared, have been closed or replaced or it has been agreed otherwise with OFCOM.

Section 5.46 is a catch-all safety net. Having dealt earlier with the specific products that were of concern to industry at the time the Undertakings were signed, Section 5.46 provides that Openreach will not supply any other product (save for various listed exceptions or as agreed with OFCOM) to other parts of BT unless it also offers that product to other Communications Providers on an EOI basis.

Section 5.47 deals with how Openreach interacts with its customers and needs to be read in conjunction with Section 8. Openreach's customers are only Communications Providers and the intention here is that Openreach should have its own sales and account management capability and be the primary channel to market for its portfolio of products. Customers, particularly where they buy from different parts of BT but wish to maintain only one account management relationship, should have the choice to decide which part of BT they deal with, provided that they experience no disadvantage in terms of price, service or quality as a result of being account managed from outside Openreach.

Section 5.49 deals with space and power in BT's exchanges for purposes related to the supply of LLU services. Concerns had been expressed as to how space was being allocated, in particular that some Communications Providers were 'bagging' space in advance of actually needing it. Openreach has developed an EOI space-only-allocation product enabling Communications Providers to pre-book space they will require for their LLU and backhaul services. In return for an appropriate charge they can pre-book space up to 18 months in advance. Section 5.49 sets out how the product will be developed, how the process will operate and be monitored, how Communications Providers can raise concerns if their demands are not being met, and how conflicts can be resolved.

In June 2009, by Variation 19, new Sections 5.51 to 5.60 in relation to 'Next Generation Access': the provision of super-fast broadband, were added.

- Section 5.51 allows Openreach to control the electronics in the access and backhaul network which are used in the provision of a 'BT Active FTTC Product'[33].

- Sections 5.52 and 5.53 define how the BT Active FTTC Product will be provided.

- Sections 5.54 and 5.55 set out how BT will develop its products and will consult with industry in that regard.

- Sections 5.56 and 5.57 deal with the provision of FTTC Passive Inputs[34] [35].

- Sections 5.58 to 5.60 set out mechanisms for reviewing the application of these commitments.

5.4.5 Section 6—BT Wholesale

Whilst the prime focus of the Undertakings, in terms of organisational separation, was on Openreach, the Strategic Review had revealed Communication Provider concerns about the development of wholesale broadband access services. To address these, OFCOM required BT to put in place 'lighter touch' commitments within BT Wholesale in terms of product management, information sharing and account management. This sector of the marketplace has however seen significant change since then with the result that Section 6 of the Undertakings has significantly evolved.

---

33 An Ethernet based Bitstream Network Access product offered by Openreach and provided over a fibre to the cabinet network.

34 The provision of copper wires, street cabinets and associated components of the Physical Layer of BT's Access network used for FTTC services.

35 Given that the set of markets in which market failure was found was:

'the upstream markets for the provision of access and backhaul network services and all the related downstream markets, including retail markets, for which the former services are a critical input.'

it is interesting to consider the appropriateness of commitments in relation to passive inputs. These might be considered to be products which are in a market which is upstream of the market in which market failure was found.

'The Undertakings'

In the 2004 Wholesale Broadband Access Market Review, OFCOM had determined that the BT product IPStream was in a market downstream of its DataStream product. It had imposed SMP obligations on DataStream and on the basis that these would suffice, BT was found not to have SMP in the supply of IPStream.

By 2005, however, most ISPs were still buying IPStream and there were concerns that BT was favouring its downstream Retail business in the supply of IPStream. For this reason, OFCOM requested, and BT agreed, that IPStream should be an EOI product—which itself in turn consumed an Openreach EOI input. So DataStream was an SMP product, but with no EOI, whilst IPStream became an EOI product, but with no SMP.

Two other areas of concern had been identified. One was in relation to retail leased lines purchased by Communications Providers. OFCOM's view was that Communications Providers wanted a variant of these with less retail service wrap and which could therefore be made available more cheaply. The other was in relation to the supply of Wholesale Calls, which again was not an SMP product.

The Undertakings required BT to partition its Wholesale product management unit.

- BTWS (for trading purposes called Core Network Services ('CNS') by BT) was made responsible for BT's wholesale SMP products[36].

- BTS (for trading purposes called Value-added Network Services ('VNS') by BT) was made responsible for a set of non-SMP products 'of significance' to Communications Providers. These were IPStream, Wholesale Calls and the proposed Communications Provider variants of retail leased lines.

- BT was permitted to product manage its other unregulated products either in a separate unit (Unregulated Network Services ('UNS')) or they could be included in either CNS or VNS.

Even though IPStream became an EOI product, there remained concerns because the variant of IPStream consumed by BT was different from the variant consumed by the majority of other Communication Providers. To address this, BT agreed to create a new variant of IPStream called IPstream Connect. This would be supplied by BT on an EOI basis and used as an input both to BT's IPStream offering and by other Communication Providers who wished to create their own IPStream equivalent service.

By Variation 14 to the Undertakings, it was agreed that when BT started to supply IPstream Connect, it would be supplied on an EOI basis—and the RFS date was set as 31 October 2008. The IBMC date is 31 March 2009. On attaining IBMC for IPstream Connect, the obligation to supply IPStream as an EOI product lapses[37].

In the 2008 Wholesale Broadband Access Market Review OFCOM reversed its previous findings in relation to product market definition and found IPStream and DataStream to be in the same product market. It then defined three separate

---

36   All wholesale SMP services excluding those provided by Openreach
37   The same Variation included provisions related to BT's Next Generation Network wholesale broadband access product Wholesale Broadband Connect, but by Variation 17 this was removed on the basis that it was not needed (having regard to the provisions of Sections 11.7 and 11.8) and the definition of Wholesale Broadband Connect was removed by Variation 18.

geographic markets, finding that BT had SMP in Markets 1 and 2 but not in Market 3 areas, which as at 2008 covered approximately 70 per cent of households.

As a consequence of the finding that BT had SMP in relation to the supply of IPStream and IPstream Connect in Markets 1 and 2, and hence that it was an SMP Product, OFCOM agreed that they could both be product managed from CNS. This has meant the 'reunification' into one place of the product management of all wholesale broadband access products.

On the grounds that it had no SMP in Market 3 areas, BT submitted that it should not be subject to Undertakings obligations, in particular any EOI obligation, in those areas. OFCOM's view, however, was that the EOI obligation imposed by Section 3 of the Undertakings (as amended to include IPstream Connect) was not dependent on a finding of SMP and hence that the EOI obligation did not automatically fall with the removal of SMP. However, by Variation 17, in December 2008, the Undertakings were amended to remove the EOI obligations in Market 3 areas.

In relation to the development of Communication Provider variants of retail leased lines, the process was as set out in Annex 3 and required BT to consult Communications Providers on their requirements. BT did so, but found there was no demand for such products, although some minor product enhancements were taken forward via BT's usual product development processes. At the time of writing there are no such products.

As a result, the only significant product still required to be product managed separately from VNS is Wholesale Calls. This is a non-SMP product, and it is anticipated that this Undertaking will be considered by OFCOM either in the course of the narrowband market reviews which commenced in March 2009 or immediately thereafter.

The table below references the different product management units and what they provide.

Undertakings Definition	BT Wholesale Trading Name	Services to be supplied as at September 2005	Services supplied as at April 2009
BTWS	Core Network Services (CNS)	Wholesale SMP products (excluding Openreach wholesale services)	Includes all BT Wholesale products other than Wholesale Calls.
BTS	Value-added Network Services (VNS)	1. IPstream 2. Wholesale Calls 3. Annex 3 leased lines. These were the non-SMP products considered to be 'products of significance to other Communications Providers'.	1. IPstream moved to CNS following Wholesale Broadband Access Market Review 2. Wholesale Calls still supplied from VNS 3. No such leased lines have been developed.
N/A	Unregulated Network Services (UNS)	Other non-regulated products. The Undertakings did not separately define them. By Section 6.3 they can be managed from within CNS or VNS.	

Figure 5.2

'The Undertakings'

Against that backdrop, the provisions of Section 6 can be reviewed.

Section 6.1 required BT to set up the separate product management units and defined the products to be included in each—SMP products in BTWS/CNS and Significant products in BTS/VNS. Section 6.3 confirmed that other products could be managed from whichever part of BT Wholesale that BT wished and Section 6.4 sets out the processes for amending the product sets in the different units.

Section 6.2 required a senior BT manager to be responsible for these units. Section 6.5 defines the requirements of product management and Sections 6.6–6.8 deal with incentive remuneration for the senior manager and CNS people.

Recognising the importance of separating BT Wholesale CNS people from retail people, Section 6.9 prohibits them from also working for a downstream business without permission from OFCOM.

The information sharing obligations in relation to BT Wholesale are set out in Section 6.10. They were amended in December 2007 by Variation 12[38] to reflect the creation of the two BT internal supplier divisions—BT Design and BT Operate. Where the Openreach information sharing requirements are predominantly designed to stop Openreach CI and CCI going outside Openreach, the Wholesale rules are less onerous and are predominantly designed to restrict information passing inappropriately to BT's downstream business divisions—BT Retail and BT Global Services.

BT Wholesale CCI may not be shared with the downstream business divisions or Openreach, nor to BT Design or BT Operate people whose incentive remuneration reflects the objectives of the downstream lines of business. The exceptions to this are with customer consent, where needed for orders to be processed, or to Annex 2 people.

Section 6.10.2 contains a provision (subject to the usual exceptions) restricting the supply of CNS CCI to VNS people, and Section 6.10.4 restricted the ability of VNS people to influence the Commercial Policy of CNS people. Save in relation to Wholesale Calls, these requirements are now spent.

Section 6.10.3 restricts the supply of BT Wholesale Commercial Information to the downstream business divisions, unless the disclosure is of the nature that would be made to other Communications Providers in the ordinary course of business, or it is to Annex 2 people.

Sections 6.11 and 6.12 have consequential provisions with regard to ensuring that BT Wholesale undertakes product development in a fair way. Section 6.13 deals with systems separation for BT Wholesale systems and Section 6.14 allows BT Wholesale to draw support from centres of excellence that sit in other parts of BT, for example the billing centre of excellence, provided that information sharing rules are respected.

Finally, Section 6.15 provides a sunset clause for the disapplication of the Section 6 obligations where a wholesale SMP product is subsequently found not to have SMP, unless BT and OFCOM agree that it should become a Significant BTS/VNS product.

---

38   http://www.ofcom.org.uk/telecoms/btundertakings/exemptionsandvariations/var12.pdf.

5.4.6 Section 7—Equipment Location

The Section 7 provisions in relation to equipment location need to be read in conjunction with the definition of Equipment in Section 2, the obligation at Section 5.19 to provide space in exchanges and the list of equipment in Annex 4.

The intention of Section 7 is to permit, where feasible, the location of Communications Provider equipment in BT's exchanges for the purpose of providing services over the access network other than Broadband and telephony. Annex 4 refers, for example, to video servers, aggregation equipment for backhaul services and private circuit termination equipment.

Section 7 sets out a process for BT to publish annually details of the space which it anticipates will be available in its exchanges. BT may invite Communications Providers to submit their observations, including their own demand forecasts for space.

Pursuant to Section 7, BT has also put in place a process for Communications Providers to occupy available space in a Communications Provider Operational Area in an exchange for their Equipment. The Communications Provider must submit a request for space, and there must be sufficient space available and sufficient electrical power. If the space requested is not available, BT may provide an Alternative Communications Provider Operational Area.

Sections 7.4 to 7.6 deal with the commercial basis on which occupation may take place. They provide that occupation will be 'on reasonable commercial terms', which include safeguarding the operational integrity of the exchange, security, and health and safety requirements. These are property related obligations that apply to BT as a whole as opposed to Openreach and by Section 7.4.4. if issues arise between BT and the Communications Provider and are unresolved, the escalation route for a Communications Provider is to a BT Wholesale Executive Board member.

Sections 7.7 and 7.8 deal with the closure of exchanges. As part of its drive for operational efficiency and allied with the building of its Next Generation Network, BT has an exchange closure programme. BT has established a process to keep Communications Providers informed of exchanges to be closed and these Sections provide a way of managing the closure process whilst safeguarding the interests of Communications Providers.

It should be noted that Section 7 does not apply to the collocation of LLU services. This is dealt with under Section 5, in particular Section 5.49.

5.4.7 Section 11—Next Generation Networks

When the Undertakings were signed, the form that BT's NGN would take was largely unknown, but industry interest in it was high. This Section of the Undertakings was therefore written in a way that sought to anticipate future events, providing safeguards so that any major change to network infrastructure did not harm the legitimate business interests of BT's competitors.

Sections 11.1 to 11.4 confirm that the qualities of the regime applying to BT's then current generation access network are to be maintained going forward. Sections

11.1 and 11.2 require BT to provide Network Access using its NGN in markets in which it has SMP, to do so on an unbundled basis and on terms that allow other Communications Providers to compete effectively. Arguably, this is no more than SMP obligations require anyway.

Sections 11.3 and 11.4 require BT to consult with Communications Providers before making network design decisions which might have the effect of preventing Communications Providers from having Network Access. If demand for a service exists, BT must go through a negotiation process before implementing its decision.

The purpose of Section 11.5 is to drive efficient network design and to safeguard Communications Providers from bearing the cost of any network design inefficiencies. It applies where BT is subject to SMP cost orientation obligations and requires that in such circumstances BT shall set its charges on the basis of efficient design, subject to a proviso which applies when a consultation on network design has been undertaken under Section 11.3.

Section 11.6 requires BT to 'build in' EOI into the design of its NGN and Section 11.7 requires BT to provide Network Access services provided by means of its NGN on an EOI basis. By Section 11.8, this obligation applies only where BT has (or is likely to be found to have) SMP in the supply of those Network Access services. Section 11.9 sets out other exceptions to the EOI obligation, including reasonable practicability and agreement from OFCOM.

The inclusion of EOI obligations subject to the existence of SMP reflected the anticipation that by the time that BT had built its NGN, Openreach would be operating the economic bottleneck services, be trading at an equivalent level and that by then SMP would be largely concentrated on Openreach.

Sections 11.10 and 11.11 are concerned with ensuring that where BT has SMP and it launches new (downstream) services which are based on NGN Network Access services, the NGN Network Access service is made available to other Communications Providers in sufficient time for them to build competing downstream services. Arguably, again, this overlaps with SMP prohibitions on undue discrimination.

Sections 11.12 to 11.14 are concerned with consultation on network design and the establishment of an industry group to allow this. BT has used its 'Consult 21' process for this purpose. Sections 11.15 to 11.17 deal with the establishment of an operational dispute adjudicator to deal with NGN related disputes. The action to create dispute adjudication lies with OFCOM rather than BT. So far OFCOM has not opted to build an adjudication regime.

It was recognised that in building its NGN BT might have to make network re-arrangements that would impact on other Communications Providers. Section 11.18 sets out possible bases for the payment of compensation in such circumstances, but they are heavily caveated and are merely 'the principles BT will use'. They do not affect any statutory, common law or contractual rights that a Communications Provider may have.

Section 11.20 allows the sharing of information across BT when otherwise the information sharing rules might prevent this, if the sharing of the information is for the purpose of enabling BT to design build and implement its NGN or if it relates to the decision making process relating thereto. Designing, building and implementing the new network is distinct from the creation of wholly new services to be run over the new network that become possible once the new network exists. For the latter, the usual information sharing rules apply.

5.5 The Legal Framework for the Undertakings

5.5.1 The Making of the Undertakings

The recitals to the Undertakings set out in brief their legal basis:

> *'WHEREAS:*
> *OFCOM considers that it has the power to make a reference to the Competition Commission under Section 131 of the Enterprise Act 2002;*
> *BT has offered undertakings in accordance with Section 154 of the Enterprise Act 2002; and*
> *OFCOM, instead of making a reference to the Competition Commission, has decided to accept BT's undertakings.'*

Section 131 authorises OFCOM[39] to make a reference to the Competition Commission ('the Commission') if OFCOM:

> *'. . . has reasonable grounds for suspecting that any feature, or combination of features, of a market in the United Kingdom for goods or services prevents, restricts or distorts competition in connection with the supply or acquisition of any goods or services in the United Kingdom or a part of the United Kingdom.'*

If a reference is so made, then the Commission, on the market investigation reference, is required to decide whether any feature or combination of features of each relevant market prevents, restricts or distorts competition in connection with the supply or acquisition of any goods or services in the United Kingdom or a part of the United Kingdom[40].

If the Commission finds that there is an adverse effect on competition, it shall then decide whether it should take action for the purpose of remedying, mitigating or preventing the adverse effect on competition concerned, or any detrimental effect on customers so far as it has resulted from, or may be expected to result from, the adverse effect on competition. It will also decide whether it should recommend that others take action, and in either case, if action should be taken, what action that should be and what is to be remedied, mitigated or prevented.[41] The time limit for preparing and publishing its report is two years from the date of the reference, unless amended by order of the Secretary of State[42].

Section 138 of the Enterprise Act imposes a duty on the Commission to take action to remedy, mitigate or prevent the adverse effect on competition or the detrimental

---

39   Throughout the Enterprise Act 2002 reference is made to the OFT. By virtue of Section 369 and 370 Communications Act 2003, the OFT's functions under Part 4 Enterprise Act 2002 apply concurrently to OFCOM. For ease of reading, whilst in the references below the Enterprise Act refers to the OFT, we shall refer to OFCOM.
40   Section 134 (1) Enterprise Act 2002.
41   Section 134 (4) Enterprise Act 2002.
42   Section 137 Enterprise Act 2002.

effects on customers as described above and the Commission is empowered to accept final Undertakings or to make orders. The order-making powers of the Commission are set out in Schedule 8. They are extremely broad and far reaching and include, inter alia, the following types of provisions:

- Orders prohibiting the making or performance of agreements, or the termination of them.

- Orders prohibiting the withholding from any person of any goods or services

- Orders prohibiting those supplying goods from imposing obligations or conditions to do things or not to do things on any person purchasing those goods (bundling) and orders prohibiting discrimination.

- Orders requiring a person to supply goods or services or 'to do anything which the relevant authority considers appropriate to facilitate the provision of goods or services'[43].

- Orders providing for the division of any business (whether by the sale of any part of the undertaking or assets or otherwise) or the division of any group of interconnected bodies corporate[44].

It will be seen from the above that it would have been within the power of the Commission, had BT been referred by OFCOM and appropriate findings made, to order structural separation, i.e. involving the separation and transfer of ownership of a part of the original undertaking, or organisational/functional separation, i.e. where the relevant part of the business is organised as a separate trading division of the same undertaking, or to impose behavioural and transactional requirements of the type found in the Undertakings today.

Whilst as described above, the Commission can, following a reference and investigation, accept undertakings itself as an alternative to the making of orders, Section 154 Enterprise Act 2002 also empowers OFCOM to accept undertakings as an alternative to the making of a reference to the Commission. It may accept undertakings where it considers: '. . . it has the power to make a reference under Section 131 and otherwise intends to make such a reference.'

BT's Undertakings were given pursuant to this power, thereby avoiding the need for a market investigation reference. As indicated above, OFCOM's Section 155 Notice of 30 June 2002[45] set out the features of the market which OFCOM had identified which prevented, restricted or distorted competition and why it considered that the undertakings proposed would be apt to remedy those features.

5.5.2 Variation of the Undertakings

Variations to the Undertakings are permanent changes to them. As at July 2009, there had been 19 Variations[46].

---

43 Enterprise Act 2002,Schedule 8, Paragraph 10(1).
44 Enterprise Act, Schedule 8, Paragraphs 11–14.
45 http://www.ofcom.org.uk/consult/condocs/sec155/.
46 These are all published at http://www.ofcom.org.uk/telecoms/btundertakings/.
    exemptionsandvariations/. Further variations in relation to the system separation obligations and FTTP services are expected later in 2009.

Section 18 of the Undertakings provides that BT and OFCOM may, from time to time, vary and amend the Undertakings by mutual agreement. However this provision is obviously subject to compliance with the procedural requirements of the Enterprise Act 2002.

Section 154(6) Enterprise Act 2002 permits an undertaking given in lieu of a market investigation reference to be varied or superseded by another undertaking. It does not explicitly permit the introduction of new commitments which were not the subject of the original undertakings, so OFCOM cannot treat the Undertakings as a form of 'Licence Mark II' and introduce wholly new commitments in areas of the industry not previously the subject of undertakings.

Section 162 Enterprise Act 2002 imposes a duty on OFCOM to keep under review the carrying out of enforcement undertakings and orders. This means that in the context of BT's Undertakings, it must, in particular, from time to time consider whether the Undertakings have been and are being complied with and whether by reason of any change of circumstances an undertaking is no longer appropriate and BT can be released from it or if it needs to be varied or superseded.

In practice, virtually all of the variations made to date have been to introduce commitments which supersede the original commitments; in effect, evolution of existing commitments, sometimes to add more specificity to originally broad concepts.

The relationship between SMP regulation and the Undertakings is considered in Section 5.5.4 below, but OFCOM's Variation 17, removing the EOI obligation from certain BT wholesale broadband products following the conclusion of the 2008 Wholesale Broadband Access Market Review[47], would support the contention that where markets the subject of the Undertakings are been found to be competitive there is a change of circumstances making the continuation of Undertakings commitments in that area inappropriate.

By Section 155 Enterprise Act 2002, OFCOM is required to consult for a period of not less than 15 days before making any material variations to the Undertakings. In relation to Variation 17, OFCOM stated that:

> '. . . in this instance, OFCOM does not consider that the proposed variation and exemption alter the Undertakings in a material respect. . . .We consider that the variation in relation to IPStream is not material because it reflects the findings of the Wholesale Broadband Access (WBA) market review and the resultant position created by deregulation. . .'[48].

However, as IPStream had been a Significant product, OFCOM concluded there was merit on consulting on the proposed variation, notwithstanding that there was no statutory obligation to do so.

It should be remembered that in addition to Variations, OFCOM and BT have agreed exemptions from particular requirements of the Undertakings and other minor changes to them. These are not formal changes made pursuant to the Enterprise Act 2002, but are based on mechanisms written into particular sections of the Undertakings[49].

---

47   http://www.ofcom.org.uk/consult/condocs/eoi/eoi.pdf.
48   http://www.ofcom.org.uk/consult/condocs/eoi/eoi.pdf. Paragraphs 1.12 and 1.13.
49   This degree of flexibility in the Undertakings is permitted by Paragraph 21, Schedule 8, Enterprise Act 2002.

5.5.3 Enforcement of the Undertakings

The legal framework for enforcement action and for the bringing of damages claims in the event of a breach of the Undertakings is set out in Section 167 Enterprise Act 2002.

Section 167 provides that any person to whom an undertaking relates shall have a duty to comply with it. The duty shall be owed to any person who may be affected by a contravention of the undertaking and any breach of that duty which causes such a person to sustain loss or damage shall be actionable by him.

So, in the event of a breach of the Undertakings, a third-party can bring a claim for damages in the civil courts. In any such proceedings, it shall be a defence for the person who has given the undertakings to show that he took all reasonable steps and exercised all due diligence to avoid contravening the undertaking.

Section 167 also empowers OFCOM to secure compliance with any undertaking which has been breached by civil proceedings for an injunction or for any other appropriate relief or remedy.

When the Undertakings were being drafted in 2005, BT and OFCOM both recognised that the use of the formal Enterprise Act 2002 power to apply to the High Court for an injunction would be a heavy-handed way of remedying what might be fairly minor issues. In consequence, a lighter touch enforcement mechanism was written into Section 15 of the Undertakings. The Section 15 process, which enables OFCOM, to make directions to secure compliance, is not dissimilar to the one set out in the Communications Act 2003 whereby OFCOM can direct a communications provider to remedy a breach of a regulatory condition.

Where OFCOM has reasonable grounds for believing that BT has breached any of the Undertakings it may give BT a written notice specifying the undertaking or undertakings concerned and its reasons and a draft direction which may specify or describe steps to be taken by BT for the purpose of securing compliance with the undertakings concerned.

BT then has a reasonable period, being a period of at least one month, to make representations to OFCOM.

If having considered those representations, OFCOM is satisfied that BT is in breach of one or more of the undertakings, it can give BT a direction as described above.

BT then has two weeks in which to notify OFCOM whether it accepts the direction or not. BT may only decline to accept the direction following a decision of the BT Group plc Board to that effect.

If BT was to decline a direction, OFCOM would then have to consider whether to exercise its formal powers of enforcement under Section 167 Enterprise Act 2002.

A failure to comply with a direction would of itself be a breach of the Undertakings.

Table 1 below provides a flow chart description of the Undertakings enforcement mechanisms.

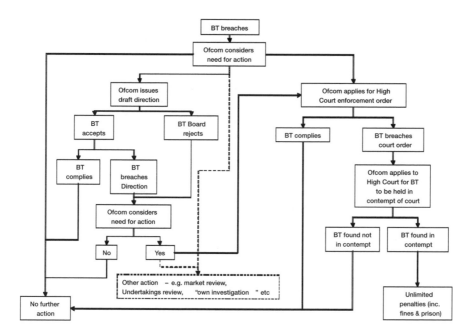

Figure 5.3

5.5.4 Revocation of the Undertakings

The Undertakings continue until such time as BT is released from them by OFCOM[50].

Given that undertakings are given for the purpose of remedying, mitigating or preventing any adverse effect on competition or any detrimental effect on customers that has resulted from an adverse effect on competition, the litmus test for revocation is, therefore, whether the adverse or detrimental effect still exists (or would continue in the absence of the undertakings). If it does not or would not, then the undertakings should be revoked. In practice we are seeing that as competition develops particular aspects of the undertakings can be revoked by Variation[51].

In addition to the Enterprise Act procedure, Section 19 of the Undertakings contains three triggers for disapplication of the Undertakings in whole or in part.

- The Undertakings automatically terminate if a reference is made to the Competition Commission under the Enterprise Act 2002[52].

- If BT is found not to have SMP in any Network Access market in the UK the entirety of the Undertakings will not apply. If BT ceases to have SMP in a particular geographical area in any network access market, the entirety of the

---

50 Section 154 (6) Enterprise Act 2002.
51 For example, EOI for IPStream and IPstream Connect was disapplied by Variation 17 in the market identified in the Wholesale Broadband Access Market Review as Market 3, which was found to be competitively supplied.
52 Undertakings Section 19.1

Undertakings will not apply in that geographical area[53]. So, when the last SMP obligation on BT in any particular area was removed, the Undertakings would automatically cease to apply in that area.

- If a BT product is supplied in a market in which BT is found not to have SMP, then any Undertakings commitments cease to apply to that product[54]. OFCOM's view, however, is that the imposition of an EOI requirement is separate from other Undertakings commitments and so the revocation of SMP for a particular product does not automatically lead to a revocation of that product's EOI obligation, even if all other Undertakings obligations in relation to it fell (e.g. information sharing obligations etc). OFCOM has however already removed the EOI obligation for wholesale broadband products in competitive geographic markets. It remains to be seen what line OFCOM will take as a result of deregulation of certain AISBO services in the Business Connectivity market review[55].

5.5.5 Appealing Undertakings Related Decisions

Section 179 Enterprise Act 2002 provides that any person aggrieved by a decision of OFCOM in connection with a reference or a possible reference under Part 4 of the Act may apply to the Competition Appeal Tribunal for a review of that decision. In determining such an application the Competition Appeal Tribunal shall apply the same principles that would be applied by a court on an application for judicial review.

Decisions relating to the acceptance, variation and release of undertakings given in lieu of a reference are considered to be decisions in connection with a possible reference and hence susceptible to judicial review pursuant to Section 179.

53 Undertakings Section 19.3.
54 Undertakings Section 19.2.
55 See Section 5.4.2 (final paragraph) above.

Chapter 6

Content Regulation

Ingrid Silver
Denton Wilde Sapte LLP

DentonWildeSapte...

Introduction

Content regulation is not a new concept. Throughout the ages, governments have sought to regulate what we see and read. From the regulation of the theatre in Shakespearian times and the controversy surrounding the publication of D.H Lawrence's 'Lady Chatterley's Lover' in 1960 to the creation of OFCOM in 2002, content regulation has had to evolve with changing social norms and the format of the content we consume.

In Western democracies at least, the philosophy behind media regulation is two-fold: to meet audience expectations and to protect viewers. In doing this, regulators aim to ensure that there is plurality in the provision of services, that those services appeal to a wide range or proportion of the public and that they are of a high quality. Regulators also seek to ensure that members of the public are protected from harmful or offensive material and that they are shielded from unfairness or the infringement of their privacy[1].

From the date of the first regular BBC broadcasts in 1936, broadcasting has expanded greatly. We now live in a diverse media world where there are five Public Service Broadcasters (PSBs), hundreds of satellite television channels and content being delivered through a variety of formats and media, such as mobile phones and the internet.

The Audiovisual Media Services Directive 2007[2] will be implemented into English law by the end of 2009 and is the latest move by the European Commission to update legislation in the media sphere to reflect advances in technology. The Directive has important implications in the field of content regulation and the interpretation the UK government takes in implementing it into national law will affect the extent to which these implications are an issue for audiovisual media service providers in the UK. For instance, we will see far more substantial regulation of video on demand services.

---

1 OFCOM presentation 27 March 2007 – Realising Television Without Frontiers, The Future of Content Regulation.
2 The AVMS Directive can be found at http://eur-lex.europa.eu/LexUriServ/LexUriServ.do?uri=OJ:L:20 07:332:0027:01:EN:HTML.

Scope of this Chapter

This chapter focuses on so-called 'traditional media', namely television broadcasting and radio as well as 'new media', which as of 2009 includes digital media diffused via the internet and wireless networks (such as 3G and Wi-Max) and on-demand video services. This chapter does not address the laws surrounding copyright, slander, defamation, libel or privacy. Laws surrounding privacy in particular will become increasingly important in a world where content can be targeted at smaller numbers of people based on their specific personal information.

The chapter will begin by plotting a brief history of how media regulation has evolved in the UK and then go on to review the current regulatory regime. The 1989 Television without Frontiers Directive and its 1997 revision will be analysed as a precursor to a detailed analysis of the new Audiovisual Media Services Directive 2007 which is due to be implemented in the UK by 19 December 2009.

Rapidly changing technology and business models present an increasing number of challenges and conundrums for content regulation. Some of these are also considered.

A Plotted History of How Media Regulation Has Evolved

The radio was invented at the end of the 19th century and marked the start of substantial developments over the following hundred years in broadcasting. By 1904, the British Government had introduced the Wireless Telegraphy Act 1904, bringing 'wireless telegraphy' under public control and requiring that all wireless operators were to have a licence. The increasing popularity of the medium lead to the creation of the British Broadcasting Company in 1922 and the subsequent constitution by Royal Charter of the British Broadcasting Corporation (BBC) in 1927.

The BBC's general manager in the 1920s wanted the broadcaster to 'educate, inform and entertain the whole nation, free from political interference and commercial pressure'.[3] The BBC's monopoly on public broadcasting in the UK ended in 1954 with the introduction of commercial television broadcasting in the form of ITV (Independent Television). There were concerns across the UK as to the effect that the ability to advertise on television and the subsequent competition between channels for viewers would have on the moral standards of television. Television in America at that time carried advertising and reports from across the Atlantic noted that 'competition has ... driven advertisers to play down what they believe is majority taste for crime, cheap sex, appeals to avarice and worse'[4]. The Independent Television Authority (ITA), a public body accountable to Parliament and created in the same year was, as part of its role, tasked with the setting up of the ITV and supervising its programmes and advertising. Following Government consultation, the ITA's supervisory role was bolstered by the Television Act 1963

---

3   The BBC Story - History of the BBC—www.bbc.co.uk.
4   A.Briggs, 'The History of Broadcasting in the United Kingdom, Volume IV, Sound & Vision' (1979) p. 894 Gerald Cock, First Director of BBC Television writing from San Francisco, *Manchester Guardian* 11 May 1953.

which required it to create a code for programme makers setting out standards and good practices to ensure good taste and decency were adhered to. Whilst the ITA did not supervise the BBC, the BBC nevertheless agreed to follow similar standards to those set out by the ITA. The ITA also gained greater control over advertising.

Whilst television was gaining popularity in the 1960s, radio was in decline and the BBC's radio programming was threatened by pirate radio stations such as Radio Caroline which were broadcasting illegally and were moored on ships just outside British territorial waters. The Government introduced legislation in 1967 to attempt to combat the pirate radio stations by making it illegal to advertise on the stations, publicise, maintain or supply them[5].

The late 1960s lead to the rise of public pressure groups, one of which was famously spearheaded by Mary Whitehouse who's campaign was entitled 'Clean Up TV'. This lobbying urged for increased control and supervision of television programmes. In the early 1970s, the ITA (by then renamed the Independent Broadcast Authority (IBA)) subjected newly introduced commercial radio to similar rules as television. The BBC continued to regulate its own radio and television services.

The rule of Margaret Thatcher's Conservative government brought about several changes in content regulation and there were significant developments in the broadcasting industry throughout the 1980s. A fourth terrestrial television channel was created, satellite and cable television were introduced, seeing rapid growth, and radio also expanded through local and community radio stations. The advent of cable television lead to the creation of the Cable Authority in 1984[6] which had responsibility for overseeing, appointing and regulating cable television services. The Video Recordings Act 1984 extended the British Board of Film Classification's (the BBFC) power to the classification of video recordings. This lead to a two tier level of censorship, with films that were watched at home (eg on video) being subject to a different level of regulation than those shown at the cinema or broadcast on television. Thatcher's strong moral stance on the display of sex and violence on television played a key part in her manifesto for the Conservative party's 1987 re-election. This strong moral stance was coupled with a desire to increase choice for the public and enable broadcasters to make the most of new technology.

The Broadcasting Standards Council (BSC) was created in 1988 to supervise programme standards, implement a code of practice and deal with complaints. The late 1980s also saw several sources of conflict between broadcasters and the government over the coverage of issues in Northern Ireland. The government gave directions to the BBC and the IBA restricting their ability to broadcast certain items and interviews regarding the Northern Ireland conflict.

The 1990 Broadcasting Act

This legislation created one of the pre-cursors to OFCOM—the Independent Television Commission (ITC)—amalgamating the IBA and the Cable Authority. The ITC's ambit was to provide licensing and 'light touch' regulation for all

---

5   Marine, &c., Broadcasting (Offences) Act 1967.
6   Cable and Broadcasting Act 1984.

commercial television services (cable, satellite and terrestrial), providing a lower level of regulation to allow the industry to grow and embrace new platforms and technologies without being hindered by the restrictiveness of traditional methods of regulation. The provisions of the Television Without Frontiers Directive 1989 (see below for detailed discussion) were also incorporated into the Act.

One of the outcomes of the Whitehouse lobbying was that the Thatcher government saw fit to apply the Obscene Publications Act 1957 to radio and television in the Broadcasting Act 1990[7].

Whereas the IBA had been able to preview programmes and schedules before they were broadcast, following the 1990 Act, the IBA's pre-censorship came to an end and the ITC would only intervene after the event. It had several powers, such as the ability to impose fines, issue warnings and ultimately remove a broadcaster's licence[8]. The ITC was charged with the creation of a Programme Code setting out details of the guidelines licensees should follow[9]. The Code placed importance on the idea of the watershed time of 9pm, after which content could include progressively more adult material, for example a higher sexual content or increased bad language. The Code provided for a relaxation of the watershed time to 8pm for certain encrypted or subscription based cable channels. Specific guidelines were provided as to the portrayal of violence, the use of bad language and the broadcasting of sex and nudity. There were also guidelines as to the impartiality of programming relating to political or industrial controversy. This Code formed a pre-cursor to the current OFCOM Broadcasting Code.

The Broadcasting Act 1990 also created the Radio Authority to license and regulate all commercial radio services (national, local, cable, satellite and restricted services) on both digital and analogue platforms. It was responsible for awarding licences with an aim of expanding the choice for listeners, managing frequencies (since frequency is a scarce resource) and regulating programming and advertising. Under both the 1990 Broadcasting Act and the subsequent 1996 Broadcasting Act, the Radio Authority had a duty to produce codes setting standards and practices to which licence holders would have to adhere as well as rules regarding ownership of licences.

The Broadcasting Act 1996

This Act established a regulatory framework for the development of digital terrestrial radio and television broadcasting. The Broadcasting Standards Council and the Broadcasting Complaints Commission were unified by the 1996 Broadcasting Act and renamed as the Broadcasting Standards Commission (confusingly, also the BSC). The new BSC was charged with the creation of a code giving advice on the prevention of unjust or unfair treatment, the protection from unwanted privacy infringement and practices to be followed in connection with the portrayal of violence and sexual content as well as standards of taste and decency generally in programmes. This code covered licensed programmes and also programmes broadcast by the BBC.

7   Broadcasting Act 1990, section 162 and Sch 15.
8   Broadcasting Act 1990, section 6.
9   Broadcasting Act 1990, section 7.

The Television without Frontiers (TVWF) Directive[10]

This Directive was adopted in 1989 as a result of developments in television and radio broadcasting technology, such as the introduction of satellite television, throughout the preceding decade. When the TVWF Directive was first implemented it was hailed by the European Commission as a 'win-win situation' whereby broadcasters could attract bigger audiences and viewers could get a greater choice of channels. The EU had decided to create a set of standards that would be applicable to all Member States in order to facilitate the free movement of television services within the EU and provide an environment in which EU broadcasters could compete more effectively with broadcasters from around the world. The Directive covered 'communication of programmes between undertakings with a view to their being relayed to the public.' It did not cover any communications services providing items of information or other messages on individual demand.

The TVWF Directive was supplemented by a second 'Television without Frontiers' Directive[11] in 1997 which updated the original Directive to take into account the advent of digital satellite television. The changes in the 1997 update were incorporated into English law by the Broadcasting Regulations 1998[12].

Key Principles of the TVWF Directive

The cornerstones of the TVWF Directive were the principles of Country of Origin and Freedom of Reception. These brought the principles of free movement of services to the broadcasting domain.

Country of Origin: The TVWF Directive established the principle that a broadcaster only comes under the jurisdiction of the regulator in its own Member State even if the service is made available in other Member States. This policy ensures legal certainty for service providers. The location of the media service provider's offices and where editorial decisions are made is used to establish the Member State in which the broadcaster is based. It may also be relevant to look at the location of the workforce and the Member State in which the broadcaster first began operating. In the case of *Commission of the European Communities v UK*[13] the ECJ held that Member States should:

> 'interpret the criterion of establishment as referring to the place in which a broadcaster had the centre of its activities, in particular the place where decisions concerning programme policy were taken and the programmes to be broadcast were finally put together.'

Freedom of Reception: The general principle indicates that Member States are to ensure freedom of reception and should not restrict retransmission on their territory of television broadcasts from other Member States. An exception was carved out whereby a Member State could provisionally derogate from the general principle if a television broadcast coming from another Member State contained

10   Directive 89/552/EC.
11   Directive 97/36/EC.
12   The Television Broadcasting Regulations 1998.
13   [1996] E.C.R.I. 4025.

any incitement to hatred based on race, sex, religion or nationality or if it contained material which may seriously harm children (unless minors were protected from such material either by scheduling time or a technical measure eg PIN codes). This could only be restricted after a series of steps had been taken. The Member State concerned must have notified the broadcaster and the European Commission in writing of the alleged infringements and its intentions if the infringements were to reoccur. The principle of freedom of reception was upheld in the case of *Paul Denuit (Case C-14/96)*[14] where the ECJ ruled that the Belgian Government could not prevent a local cable television operator from re-transmitting:

> *'on its territory ... the television broadcasts of a broadcaster under the jurisdiction of another Member State on the grounds that the broadcasts did not conform with the Articles 4 and 5'*

(of the TVWF Directive, regarding the promotion of European works). The ECJ ruled that it was:

> *'solely for the Member State from which the broadcasts emanated to ensure compliance with the directive and the receiving Member State was not authorised to exercise its own control'*

except in exceptional circumstances as prescribed by Article 2a(2) and detailed above.

Events of Major Importance: The TVWF Directive further attempted to protect the public by providing that Member States were *allowed to* take measures to ensure that events which are 'regarded by that Member State as being of major importance for society' were not broadcast exclusively by a broadcaster so that a substantial proportion of the public in that Member State was unable to follow the events either by live or deferred coverage on free television.

If Member States decided to regulate important events in such a manner, they had to decide on a list of events which they deemed to be important for society and decide in what format (live/partially live) they were to be broadcast in. Member States had to inform the European Commission of any action taken under this paragraph. Member States also had certain obligations to ensure that if a broadcaster in their State had exclusive rights, they did not exercise those rights in a way which deprived a substantial proportion of the public in another Member State from following those events either wholly or partially live, if that other Member State had designated the event as one of major importance for society. This was supported in the case of *R. v Independent Television Commission ex.p TVDanmark 1 Ltd*[15] where the House of Lords upheld the ITC's decision to refuse a UK based broadcaster, which had purchased exclusive rights to the Danish national football team's away football matches, consent to exercise broadcast rights. The matches were to be broadcast to Denmark where they had been designated events of national importance. The broadcaster was only able to broadcast to 60 per cent of the Danish population and refused to comply with the Danish requirement that rights obtained by a non-qualifying broadcaster (offering coverage to less than 90 per cent of the population) should be offered to other qualifying broadcasters at a reasonable price.

14   [1997] 3 C.M.L.R. 943.
15   [2001] W.L.R 1604.

Quotas: The production and access to European works were to be promoted through applying quotas. 'European works' were defined quite widely. This promotion could be in the form of financial contribution to the production of such works, a minimum of 10 per cent of their programming budget, or the share of the transmission time which was taken up by such works, again 10 per cent (excluding certain types of programme such as news, sports and advertising). Member States had to report to the European Commission every two years on the implementation of this initiative.

Protection of Minors: The TVWF Directive required measures to be implemented which prevented programmes from seriously impairing the development of minors, especially those including pornography or gratuitous violence. The Directive also required that other programmes that would have a similarly detrimental effect on the development of minors should either be broadcast at a time when minors would not normally see or hear them or technical measures should be used to prevent this. Furthermore, un-encoded broadcasts of such programmes should be preceded by an acoustic warning or identified by a visual symbol throughout their duration.

Advertisements: The TVWF Directive required that television advertising and teleshopping be distinguishable from other parts of a programme by visual and/or aural means. Both surreptitious (advertising which might mislead the public as to its nature) and subliminal advertising were prohibited. There was a prohibition on advertisements prejudicing respect for human dignity, including any discrimination based on sex, racial origin, nationality, and those that are offensive to religious or political beliefs. Advertisements could not encourage behaviour that could affect health and safety or prejudice the environment. All advertisements for cigarettes and tobacco products were prohibited. Alcohol advertising could not be aimed at minors and, amongst other restrictions, could not encourage excessive consumption. The TVWF Directive also imposed a blanket prohibition on advertising for prescription-only medical products and treatment. There were further restrictions on advertisements not causing moral or physical detriment to minors.

Rules regarding the insertion of advertisements in programmes ensured that the integrity and value of the programme was maintained, for example in sports programmes they were only to appear in the intervals in the game. Audiovisual works such as feature films and films made for television could only be interrupted once for each period of 45 minutes. Further interruptions to these programmes were allowed if their scheduled running time was at least 20 minutes longer than two periods of 45 minutes. Other types of programmes could have a minimum of 20 minutes between advert breaks. No advertisements could be placed in programmes of religious service nor in news programmes, documentaries, religious programmes or children's programmes of less than 30 minutes duration. The maximum total advertising time per hour was 12 minutes under the TVWF Directive, with a maximum average of nine minutes per hour per day.

Sponsorship: The Directive made it clear that sponsorship should not influence the editorial independence of broadcasters both in terms of programme content and scheduling. The sponsorship had to be clearly identifiable by a name and logo of the sponsor at the beginning and end of the programmes and viewers were not to be encouraged to purchase or rent products or services of the sponsor. There was a ban on sponsorship by companies whose main activity was the sale of cigarettes or tobacco products. Companies who sold medicinal products and treatments could sponsor programmes but could not promote specific medicinal products or treatments available only on prescription. News and current affairs programmes could not be sponsored.

Stricter Rules:- Despite the minimum set of common rules introduced in relation to the above topics, Member States were free to require that television broadcasters over which they had jurisdiction comply with rules that were stricter than those rules laid down in the Directive.

OFCOM and the Communications Act 2003

This section provides an overview of OFCOM and the Communications Act 2003 from a content regulation point of view. For more detail on this topic generally see Chapter 3 and for more detail on the Communications Act 2003 please see Chapter 8. The Office of Communications Act 2002 established the current communications sector regulatory body, the Office of Communications (OFCOM). OFCOM inherited the duties of the five regulatory bodies it replaced: the Independent Television Commission (ITC), the Broadcasting Standards Commission (BSC), the Radio Authority, Oftel and the Radiocommunications Agency. Therefore OFCOM's responsibilities lie across the fields of telecommunications, radio, television and wireless communications. The Communications Act 2003 (CA 2003) sets out OFCOM's duties and functions. It is responsible for furthering the interest of citizens in communications matters and where appropriate, promoting competition in relevant markets[16]. OFCOM covers both infrastructure and content in this sector and it is independent of the government. The CA 2003 sets out six main duties of OFCOM: to ensure the optimal use of the electro-magnetic spectrum; to ensure that a wide range of electronic communications services are available throughout the UK; to ensure that a wide range of high quality television and radio programmes appealing to a range of tastes and interests are available across the UK; to ensure that there is a plurality of providers; to ensure that standards are implemented to protect the public from offensive and harmful material in television and radio services; and to apply standards which protect the public from unfair treatment and invasions of privacy. For the purposes of this chapter, the focus is on the final two of these duties.

Under section 319 of the CA 2003, OFCOM is under a duty to set and review standards for the content of radio and television programmes, ensuring that they conform with the standards and objectives listed in this section of the Act. As well as a number of standards regarding advertising (such as restrictions on political advertising and sponsorship and a prohibition on surreptitious advertising) the main standards are to:

- protect persons under the age of 18;

- prevent the inclusion of any material which is likely to encourage or incite the commission of crime or disorder;

- ensure that news programmes are presented with due impartiality and adhere to the impartiality requirements set out in section 320 of the Act (Special Impartiality requirements);

16   Communications Act 2003, section 3(1).

- ensure that news is reported accurately;

- ensure that the proper degree of responsibility is exercised with respect to the content of religious programmes; and

- ensure that generally accepted standards are applied to the content of television and radio programmes so as to provide adequate protection for members of the public from the inclusion of offensive and harmful material.

The CA 2003 also prescribes certain matters to which OFCOM must have regard when setting or revising standards under section 319: the degree of harm or offence likely to be caused; the likely size and composition of the potential audience; the likely expectation of the audience as to the nature of a programme's content and the extent to which the nature of the programme's content can be brought to the audience; the likelihood of persons who are unaware of the programme's content being exposed, through their own actions, to the programme; the desirability of securing that the content of services identifies when there is a change affecting the service, in particular a change relevant to these standards; and the desirability of maintaining the independence of editorial control over programme content.

Licensing

One of the main duties which OFCOM took over from its predecessors was the licensing of television and radio services.

Under section 211 of the CA 2003 (and the preceding 1990 and 1996 Broadcasting Acts), OFCOM is required to regulate all types of television service (generally excluding those provided by the BBC). These include terrestrial television channels (digital and analogue) and also what the CA 2003 describes as 'television licensable content services'. This is defined in section 232 of the CA 2003[17] as a service broadcast from a satellite, made available through a radio multiplex or distributed through an electronic communications network, which consists of television programmes and/or electronic programme guides and is 'made available for reception by members of the public'. OFCOM has similar regulatory duties with respect to independent (i.e. not BBC run) radio stations[18].

Practical Issue—Internet Television

With the increasing rise of the distribution of television services via new platforms, such as the internet, it has been necessary to adapt how the licensing rules to these new platforms. In the case of television services provided via the internet, these services will require a licence as 'television licensable content services' (a TLCS licence) if they satisfy the definition in section 232 CA of the 2003 as set out above. They are distributed by an electronic communications network and unless restricted to a private network, they will be 'made available for reception by members of the public'. It depends on the nature of the content whether its can be considered a 'television programme' or electronic programme guide. In general, content which is transmitted in accordance with a schedule and which looks and feels like television is likely to be considered a 'television programme' and therefore require

---

17   As amended by The Television Licensable Content Services Order 2006.
18   CA 2003, section 245.

a licence (though in practice the service provider may in fact be able to rely on the TLCS licence held by the channel whose content they are transmitting rather than needing to obtain their own). However, if the television service only forms part of an electronic communications service, for example a small part of a general website, a licence may not be required in accordance with section 233(3) of the CA 2003. On-demand services (where the viewer selects which programmes to watch at a time of their choosing) are not currently regulated under the CA 2003, and although this is due to change with the introduction on the AVMS Directive there is no suggestion that they will in future require a TLCS licence.

OFCOM 's regulation of television and radio services is undertaken through the medium of licences which, following application by broadcasters, it issues to authorise them to broadcast their services. The licensed broadcasters must abide by a number of rules, including compliance with the OFCOM Broadcasting Code.

OFCOM licenses the public service broadcasters (PSBs), ITV, Channel 4 and Channel 5. Under the CA 2003 it imposes certain obligations on these broadcasters to deliver public service programming and services covering a wide range of subject matters, meeting the needs and interests of as many different audiences as practicable. The licences for these broadcasters are very comprehensive and set out the public service obligations in great detail. The obligations include requirements to meet high standards, to inform, to educate, to entertain and to reflect and support cultural activity in the UK. They should also provide a sufficient range and quantity of high quality programmes aimed at young people and children. Services should reflect the lives and concerns of different communities, cultural interests and traditions in the UK and an appropriate range of programmes should be made outside the M25 area. There are also quotas requiring services to have, amongst other things, minimum per centages of original programming, and regional programming, exact per centages of which are specified in the licences.

Whilst the licences for the three public service broadcasters are technical and difficult to obtain, licences for other television channels, such as satellite channels do not contain the PSB obligations and are much more formulaic in their allocation.

With both types of licence there are requirements to ensure that owners of licences are fit and proper. There are restrictions on local authorities, political bodies, religious bodies, advertising companies and their controlling companies, officers or associates from owning a licence or controlling a licensed company.

Radio licensing has similar requirements to the licensing of television broadcasting. OFCOM grants licenses to provide 'independent radio services' which it defines as sound broadcasting services; radio licensable content services and additional radio services[19] and it licenses a range of different radio services which are defined in section 245 CA 2003, with the exception of those provided by the BBC. OFCOM requires that radio caters for all different tastes of listeners. To achieve this, OFCOM is required to try to ensure that amongst the radio services to which it grants licences there is at least one service which consists mostly of the spoken word and one service that broadcasts music that the Broadcasting Act 1990 describes as not

---

19   Broadcasting Act 1990, section 85(1), (8).

being 'pop music'[20]. The requirements with regards ownership of radio licences are similar to those for television. In particular, following the furore surrounding pirate radio stations in the development of commercial radio in the UK, those who have been convicted of radio piracy and related offences are prevented from being able to hold a radio licence.

The Wireless Telegraphy Act 2006 (discussed in greater detail in Chapter 7) was introduced to consolidate existing legislation (including the Wireless Telegraphy Acts 1949, 1967 and 1998 and the Marine Broadcasting (Offences) Act 1967) regarding OFCOM 's management of the radio spectrum. The consolidated Act continues to reinforce the fact that it is an offence to broadcast a radio service without a licence from OFCOM. The legislation retains the strict restrictions which effectively make it an offence to provide any form of assistance to pirate radio stations.

The Broadcasting Code

Under the CA 2003 and previously the Broadcasting Act 1990, OFCOM is required to produce a code covering programme standards, fairness, sponsorship and privacy in television and radio. The Broadcasting Code gives effect to a number of the provisions of the TVWF Directive and the most recent Code was published in October 2008[21]. The Broadcasting Code contains ten sections covering the following topics: Protecting the under eighteens; Harm and Offence; Crime; Religion; Due impartiality and due accuracy and undue prominence of views and opinions; Elections and referendums; Fairness; Privacy; Sponsorship; and Commercial references and other matters. Each section sets out the principle that OFCOM aims to uphold and then details the rules that apply in order to fulfil the aim. Rules are clearly set out and easy to follow, providing clear guidance. In drafting the Code, OFCOM took into account the standards and objectives set out in the CA 2003, which would include those at section 319, detailed above. Broadcasters must ensure that they comply with the rules in the Code and there are sanctions for non-compliance.

Breach

In the event of a breach of:

> 'any content or content related requirements (including breaches of OFCOM's codes, for example the Broadcast Code) by a broadcast licensee, the BBC or S4C ('a broadcaster')'[22]

OFCOM has a number of sanctions at its disposal. It may issue a direction not to repeat a programme; require that the licence holder broadcast a correction or OFCOM 's findings on its service; impose a financial penalty; or shorten or even revoke a licence in certain cases (although revocation is not possible with respect to

---

20   Broadcasting Act 1990, section 85.
21   The Broadcast Code can be found at http://www.ofcom.org.uk/tv/ifi/codes/bcode/bcode.pdf.
22   OFCOM outline procedures for statutory sanctions in content and content-related cases, 1 February 2008.

the BBC, S4C or Channel 4). Generally, the maximum financial penalty is the greater of £25,000 or 5 per cent of the broadcaster's qualifying revenue. For the licensed PSBs, the maximum is 5 per cent of their qualifying revenue and for the BBC, or S4C it is £250,000.

OFCOM offers a right to the broadcaster to make representations regarding the breach before reaching a conclusion and imposing a sanction. The OFCOM Board has delegated the power to impose sanctions to the Content Sanctions Committee, comprising five members from both the OFCOM Board and the Content Board (a committee of the OFCOM Board with responsibility for content). Prior to the Content Sanctions Committee making its final decision, there is a rigorous procedure involving the drafting and consideration of several sanctions papers. These outline the particulars of the breach and detail the representations by the broadcaster ensuring that only cases that are of sufficient seriousness are sent for consideration by the Content Sanctions Committee, outlined in a final sanctions paper. If the Committee believes that the matter should be considered, then it invites the broadcaster to submit oral representations before then making its decision, which is final. The broadcaster is then provided with the Committee's decision shortly before it is made public. Adverse publicity is also of course a significant consequence from the broadcaster's point of view as OFCOM publishes its findings.

The Regulation of the BBC

The BBC is not regulated like other television channels in the UK and is not regulated by OFCOM in the same way. It was established by Royal Charter (and not statute), meaning that the government does not have to pass legislation through Parliament in order to set the BBC's remit and duties. The BBC has to observe certain standards set out in section 319 CA 2003 (as detailed above) with respect to its publicly funded services: the protection of under eighteens; exclusion of material likely to encourage or incite the commission of crime or disorder; responsibility with respect to religious programmes; standards regarding the inclusion of harmful and offensive material in services, fairness and privacy. The BBC's commercial services must, however, observe the entirety of the section 319 CA 2003 standards.

The BBC Trust was established by Royal Charter and its role is to act on behalf of licence fee payers to ensure the independence of the BBC and that it provides high quality output. The BBC's editorial guidelines and any proposed amendments are drawn up by the BBC's executive board, but must be approved by the BBC Trust. OFCOM agreed a formal memorandum of understanding with the BBC Trust in March 2007 governing the areas of interaction between the two bodies. According to this:

> 'the Trust is required to establish and maintain procedures for the handling and resolution of complaints about standards in the content of the BBC's services, including complaints regarding BBC editorial guidelines designed to secure appropriate standards, accuracy and impartiality, OFCOM 's Fairness Code and the relevant programme standards.'[23]

23 Memorandum of Understanding between OFCOM and the BBC Trust, p4.

The Memorandum of Understanding continues that:

> *'OFCOM is also required to consider and adjudicate on fairness complaints and to establish procedures for the handling and resolution of complaints about the observance by the BBC of standards' (under Section 319 CA 2003 as set out above).*

The BBC's Public Service remit is set out in its Charter and the Agreement. The Agreement was made between the BBC and the Secretary of State for Culture Media and Sport and was approved after a debate in Parliament in July 2006. The Agreement complements the Charter, setting out in more detail the role, rights and duties of the BBC as a public service broadcaster.

Regulation of Advertising

The Communications Act 2003 initially identified OFCOM as having responsibility for regulating advertising in broadcasting. This was delegated by OFCOM to the Advertising Standards Agency, which already regulated non-broadcast advertising. OFCOM retained control over political advertising on television, the quantity of advertising on television and the sponsorship of television programmes. The industry is therefore now co-regulated. Several Codes of Practice govern the industry and they follow the general principles that advertisements should not mislead, cause harm or offend. Two industry Committees of Advertising Practice, the CAP (Broadcast) and the CAP (Non-broadcast) are responsible for writing the Codes of Practice and the Advertising Standards Authority (ASA) administers them independently.

Committee of Advertising Practice (CAP)

The CAP is responsible for the advertising Codes of Practice in the UK. OFCOM contracts the CAP (Broadcast) to create the Codes governing television and radio advertising and enforce them. They do not cover all aspects of television broadcasting. The CAP (Broadcast) Code governs the content of television and radio advertisements which are broadcast on television channels licensed by OFCOM as well as those advertisements on interactive television services, shopping channels and Teletext services. The CAP (Non-broadcast) is responsible for the content of advertisements that are in print, on new media, on posters and at the cinema as well as sales promotions, personal data used for direct marketing, refunds and the delivery of mail order goods.

Clearcast

Broadcasters are under an obligation to ensure that their advertising is checked for compliance with the advertising Codes prior to broadcast, as part of the conditions of the licence that OFCOM grants them. An organisation called Clearcast (previously the Broadcast Advertising Clearance Centre (BACC)) is responsible for performing this function and reviewing advertisements appearing on either terrestrial or satellite channels in the UK prior to their broadcast. Clearcast ensure that the advertisements conform to the Clearcast Notes of Guidance and the CAP (Broadcast) Code.

Advertising Services Authority (ASA)

The ASA is responsible for regulating advertisements, direct marketing and sales promotions in the UK. Funding for the ASA is through a 0.1 per cent levy on advertising airtime which is collected by the Broadcast Advertising Standards Board of Finance (BASBOF) in order to ensure the ASA's impartiality. The ASA apply the advertising standards Codes (created by CAP, as detailed above) which are divided into three: television, radio and other forms of advertisement. Rules also exist to regulate advertisements on Teletext and interactive advertisements and the scheduling of television advertisements. Some products, such as alcoholic beverages and marketing techniques, also have their own individual rules. The ASA handles and resolves complaints about both broadcast and non-broadcast advertisements; ensuring that advertisers comply with the Codes as well as ensuring that the Codes are maintained and kept up to date (they advise the CAP on any amendments made to the Codes). They carry out research relating to advertising in order to ensure that the Codes reflect changing consumer views and needs as well as technological developments. They also carry out their own spot-checks on advertisements to ensure that they are compliant.

Upon receipt of a complaint, the ASA complaints team will make an initial assessment to determine whether the advertisement has breached any standards. In the event that a breach is minor, the team will often decide to resolve the problem informally and swiftly, notifying the Broadcast Council (comprised of an independent chairman and a majority of members from outside the advertising industry) of its decision. If this is not possible or the breach is more than minor, the ASA invokes a formal investigation. The ASA investigates the incident and contacts the advertiser, collecting its written responses. Following the collation of responses, the ASA case worker handling the complaint will prepare a draft recommendation setting out the findings of their investigation and a draft assessment of the outcome of the complaint. Once this draft recommendation has been sent to the complainant and the advertiser for comments on factual accuracy, it is sent to the Broadcast Council, which then makes its adjudication. The ASA case worker handling their case then informs the advertiser and complainant of the outcome and any remedial action required. Adjudication outcomes are published on the ASA's website.

An independent review procedure exists to ensure fairness. There is a compliance team to ensure that the Council's rulings are adhered to. If a complaint is upheld, the costs can be high for the offending advertiser. They will not be allowed to use the advertisement or advertising technique for future marketing initiatives and the publication of the outcome on the ASA website can lead to unwanted negative publicity. OFCOM prescribes in the licences it awards to broadcasters that they must adhere to ASA decisions. In certain instances, the ASA can decide that further sanctions are required and it will inform the broadcaster that the case will be referred to OFCOM. Broadcasters can be referred to OFCOM for persistent broadcasting of advertisements that break the rules of the Codes, which could in extreme cases lead to their licences being removed.

Other examples of co-regulation are listed below.

Self and Co-Regulatory Bodies

The Association for Television On-Demand (ATVOD)

The Association for Television on Demand was created at the time of the Communications Act 2003. ATVOD provides guidance and a Code of Practice to the self-regulating on-demand industry[24]. The Guidance notes to the Code of Practice define ATVOD's remit as covering 'all On Demand audiovisual content that is delivered through any fixed line or wireless delivery mechanism'[25]. The remit and exclusions in the latest version of the Guidelines cite the Audiovisual Media Services Directive (discussed below). As a result, ATVOD is not responsible for any user generated content or private websites.

ATVOD will act if there is an infringement of the Code by one of its members and offer an independent complaint appeals process to its members and viewers. ATVOD's complaints resolution process requires that, in the event that the service provider's responses have been unsatisfactory, a complainant first contact the offending service provider before contacting ATVOD itself. In the event that ATVOD upholds a complaint, it has a variety of remedies at its disposal. It can require the offending member to remedy the cause of the complaint as well as impose requirements regarding future conduct, reimbursement of service charges incurred by the complainant and reimbursement of administration costs incurred by ATVOD as well as the imposition of a fine and publication of the ATVOD decision and the identity of the offending member. In extreme cases, the member can even have its ATVOD membership suspended. An independent appeals adjudicator is available to hear any appeals and their decision is final and binding.

The association also provides information and advice to its members as well as the Government, OFCOM , the Department of Culture, Media and Sport and other bodies. Members of ATVOD at the time of writing include: BT, ITV, Virgin Media, Channel 4, Filmflex, Tiscali and Five. The BBC is an affiliate member and the On Demand Group is an associate member.

The Code of Practice

ATVOD's Code of Practice provides standards by which on-demand service providers can operate and is accompanied by helpful guidance notes. The Code is guided by four main principles: that

> *'children and minors should be protected from unsuitable content; all users are protected from advertising and other commercial communications which are not legal, decent, honest and truthful; adequate information about the nature of content is available before it is viewed; service providers keep their promises to users.*[26]'

The protection of minors is a key tenet of the Code and ATVOD encourages clear labelling and systems requiring user consent, such as a PIN or account holder

---

24   The ATVOD Code of Practice can be found at http://atvod.co.uk/assets/documents/general/pdf/ atvod_code_of_practice.pdf.
25   ATVOD Guidance Note No 1.
26   ATVOD Code of Conduct p1.

confirmation, to put this protection into practice. The guidance notes elaborate on rules relating to promotional material, protection of human dignity and those against harm and offence and third party content.

It is likely that ATVOD will play an important role in shaping the co-regulatory framework for on-demand services following the implementation of the Audiovisual Media Services Directive into English law in 2009.

Telephone-based Services

PhonepayPlus

The Communications Act 2003 gives OFCOM the responsibility of regulating premium rate services. This responsibility has been delegated by OFCOM to an agency. Formerly the Independent Committee for the Supervision of Standards of Telephone Information Services (ICSTIS), PhonepayPlus is the regulatory body responsible for all premium rate charged telecommunications services (PRS). PRS is defined by PhonepayPlus as including 'any premium rate goods and services that (consumers) can buy by charging the cost to (their) phone bills and pre-pay phone accounts', for example adult chat lines, mobile ringtones, TV voting and competitions. In the UK, PRS includes all numbers in the 09 number range, short codes for text messages and from August 2009 will also cover some numbers in the 087 number range, as described below. PhonepayPlus is industry-funded and consists of a board of ten members, seven of whom have no connection with the PRS industry. The three who are active in the industry do not take part in PhonepayPlus' adjudicatory functions. In fulfilling its aim of consumer protection, it regulates all aspects of PRS content, promotion and operation. The PhonepayPlus Code of Practice[27] sets out the rules by which providers of PRS must adhere. Any complaints received by PhonepayPlus are investigated and if a breach of the Code is discovered, it has the power to fine companies and bar their services. A Code Compliance Panel conducts adjudications, sitting as a formal tribunal consisting of both lay and legal members. The increase in the availability of PRS resulting from the boom in the mobile phone industry has expanded the scope of work in which PhonepayPlus is involved.

The Code of Practice

This sets out in detail the standards which PRS providers must comply with and is accompanied by non-binding help notes. The Code imposes several obligations on network operators, requiring them to perform a level of due diligence on PRS providers before allowing them to operate on their networks and specific obligations exist obliging the network operators to help PhonepayPlus in its investigations. There are sanctions for non-compliance with these obligations. Information provision obligations are also placed on PRS providers. PhonepayPlus reserves the right to request that providers of certain categories of PRS, a list of which is published from time to time, seek its prior permission before initial operation. The Code sets out detailed directions on harm and offence, pricing and promotion

---

27   The PhonepayPlus Code of Practice can be found at http://www.phonepayplus.org.uk/upload/
     PhonepayPlus_Code_of_Practice.pdf.

of services. Specific types of service also receive detailed guidelines, for example live services, dating services, competitions and other games with prizes, children's services and sexual entertainment services. The second part of the Code sets out the investigation and appeals procedures and the sanctions which PhonepayPlus can impose. Sanctions include the imposition of a fine, prevention of the person operating the offending service from running any other PRS for a specified period and a requirement for the service to refund customers.

Regulation of 087 Numbers

In the UK, non-geographic numbers starting with 0871 are frequently charged at higher rate than calls to ordinary geographic landline customers. The service provider may take a revenue share of the call charge to help pay for the service provided. 0871 numbers are a subsection of 'Number Translation Services' (NTS) and PhonepayPlus has described them as:

> *'involving a form of micropayment whereby revenue can be shared between the company you are calling and the third party that's providing the telecoms service'.*

These types of numbers are used for many different purposes, from ticket booking facilities to customer service help lines. OFCOM has also recently introduced new 0872 and 0873 numbers because capacity on the 0871 number is running low.

As part of OFCOM 's review of Number Transition Services (NTS) in April 2006, it reviewed the role of the 087 numbers and whether they should be classed as PRS. Following consultation, OFCOM announced in its Statement of February 2009 that from 1 August 2009 all 087 numbers (except 0870) which cost more than five pence per minute will be classed as PRS and will be regulated by PhonepayPlus[28]. OFCOM 's aim in including these numbers within PhonepayPlus' PRS regulatory remit is to increase consumer protection in this area and provide greater price transparency. 0870 numbersare subject to separate proposals and will not be regulated by PhonepayPlus. As a result of this change, a large number of organisations who were not previously under PhonepayPlus' remit may now be subject to their Code of Practice. Organisations providing 087 services will need to register with PhonepayPlus by 31 July 2009. PhonepayPlus is seeking to minimise concerns these companies may have by setting out detailed information about the change and the compliance obligations on its website, as well as offering compliance advice to those firms who contact them.

IMCB

The Independent Mobile Classification Body (IMCB) is responsible for setting a classification framework for certain forms of mobile commercial content. With the advent of new technologies on mobile handsets in the last decade, in 2004 the UK mobile operators all signed up to a code of practice designed to protect children from unsuitable content on mobile handsets and to promote responsible use of mobile phone services. They appointed the IMCB, a subsidiary of PhonepayPlus, to

---

28 The OFCOM Statement can be found at http://www.ofcom.org.uk/consult/condocs/087prs/ 087statement/.

classify content that is only suitable for access by those customers over the age of 18. It is a non-profit organisation funded by the mobile operators, but it retains its independence.

The IMCB determines a classification framework which mobile content service providers are then able to use, of their own accord, to identify when their content should be classified as '18' rated. Content classified as only being suitable for access by those customers aged over the age of 18 is then subject to an initial age verification check (for example a nominal credit card payment made by the customer) and what the framework describes as 'Access Controls', which include barring, PIN controlled access and subscription services to ensure that only those over the age of 18 are able to access the classified content. The IMCB's classification framework covers still pictures, video and audiovisual material, and mobile games, including java-based games. The IMCB's remit does not cover several key areas which are already governed by other bodies, for example any services which are designated as PRS services and therefore already covered by PhonepayPlus, subscriber generated content, and content which is accessed by WAP where the mobile operator merely provides connectivity. To aid content providers, the IMCB provides non-binding advice on classification.

The classification framework sets out a complaints process for consumers if they believe that an item of commercial content should have been classified as 18, as well as an appeals process run by the Classification Framework Appeals Body (CFAB). The IMCB does not deal with public enquiries. Members of the public raise issues with their mobile operator and if they are not satisfied with the response they receive, the complaint is passed to the IMCB. Importantly, the IMCB will not act as a regulator, enforcing misclassification by providers.

The Internet

ISPA

The Internet Service Providers' Association (ISPA) promotes competition, self-regulation and development of the Internet industry. It is the UK's Internet Service Provider (ISP) trade association. Membership is voluntary, but those that become members agree to abide by the rules of the ISPA Code of Practice[29].

The Code of Practice is very broad and sets out the general rules which members of the ISPA agree to follow. Its overriding theme is consumer protection and it promotes clarity of information for the consumer. The Code seeks to ensure that members adhere to the law in the promotion and provision of their services. There are conditions regarding decency and the code specifically aims to ensure that 'child abuse images' are not provided on the services provided by its members. The Code advocates principles of honesty and fair trading when dealing with customers. Specific reference is made to the observance of broadcast advertising codes administered by the ASA as well as PhonepayPlus' Code of Practice in both promotional activities and material available through member services. Members are encouraged to join the Internet Watch Foundation (IWF—see below) and co-operate with its recommendations. The Code encourages clarity in pricing

---

29   The ISPA Code of Practice can be found at http://www.ispa.org.uk/about_us/page_16.html.

and assistance in the transferral of domain name management from one ISP to another.

It sets out a detailed complaints procedure and requires all members to belong to one of the two ADR schemes approved by the ISPA Council (a body consisting of ten people from a range of different members who amongst other things decide upon ISPA policies), at the time of writing, CISAS and Otelo[30]. The complaints procedure encourages a complainant to approach the relevant member to try and sort the issue out amicably and suggests ADR as the final solution if this is not possible and a deadlock is reached. Complaints which fall under the ambit of another regulatory body, such as PhonepayPlus or OFCOM, may be referred to that body. Sanctions available include requiring a remedy of the breach; suspension of the offending member's ISPA membership and in extreme cases expulsion from the ISPA. The ISPA can publish all or part of its decision. Membership is seen as a badge of approval which consumers respect and adverse publicity and/or loss of membership would create negative publicity for the ISP.

Internet Watch Foundation (IWF), the Internet Content Rating Association (ICRA) and EU Child Protection Code

IWF[31]

The IWF is a UK organisation which works in partnership with numerous bodies, including the government, the online industry, law enforcement, charities, the education sector and the public in general. Their aim is to:

> *'minimise the availability of (potentially illegal online content), specifically, child sexual abuse content hosted anywhere in the world and criminally obscene and incitement to racial hatred content hosted in the UK.' Much of their work involves working with authorities across the world in their bid to 'encourage wider adoption of good practice in combating online child sexual abuse.'*

The body is funded by both the EU and the wider online community; however, it maintains its independence.

They operate a hotline to allow members of the public to report instances of such content but do not themselves have any ability to impose sanctions. They also provide a 'notice and take-down' service, informing internet hosting services if any such illegal content is found on their servers. In addition, they pass reports to law enforcement agencies regarding their findings of potentially illegal child abuse images online and they work closely with the police to help them in their investigations.

The ICRA[32]

The role of rating and labelling websites has been taken by the ICRA (Internet Content Rating Association); a non-profit international organisation working to

---

30 These can be found at: CISAS http://www.cisas.org.uk/ and Otelo http://www.otelo.org.uk/.
31 www.iwf.org.uk.
32 www.fosi.org/icra/.

create a safer internet. Internet content providers fill out an ICRA questionnaire detailing set vocabulary which describes the content on their site. The ICRA then creates a label from this data which can be identified by filtering software. This enables parents with such software to control the content which their children are able to access. Importantly, the ICRA does not rate the website, the content provider does. It is free to label a site with ICRA, though the ICRA provides a paid checking service which, following a check of the label by the ICRA, leads to the site being awarded an ICRA-checked label and it is added to a database of trusted and checked ICRA websites. The ICRA is part of the wider Family Online Safety Institute (FOSI).

EU Child Protection Code[33]

On 11 February 2009, seventeen social networking sites, including Facebook, Bebo and MySpace, signed a voluntary code of conduct to protect children online. The code was brokered by the European Commission. The aim is to eliminate child bullying online and to increase awareness of how children can protect their personal information. The social networking sites will include an easy-to-use and accessible 'report abuse' button on their web pages. Users registered as being under eighteen will have their privacy settings set to 'private' as default and private profiles of under eighteens will not be searchable on the web site or through search engines. The social networking sites also agreed to make it more difficult for those under thirteen to register on the sites and to ensure that the privacy options are easily accessible and available at all times.

The Future of Content Regulation

The Audiovisual Media Services (AVMS) Directive

Upon announcement of the adoption of the new Audiovisual Media Services Directive[34], Viviane Reding, the EU Commissioner for Information Society and Media said 'today the dawn of Europe's convergent audiovisual services industry is breaking'. On the 19 December 2007 the AVMS Directive became law in Europe following its approval by the European Council. The impetus behind its introduction was to meet the demands of the future and ensure legal certainty. The AVMS Directive brings in legislation requiring the regulation of on-demand audiovisual media services for the first time. Member States need to have implemented the Directive into their national law within two years (ie by the end of 2009).

Background and Reasons for Implementation

OFCOM has defined convergence as:

> *'the ability of consumers to obtain multiple services on a single platform or device or obtain any given service on multiple platforms or devices'[35].*

---

33   Details of the voluntary code can be found at http://tinyurl.com/dy6rkh.
34   2007/65/EC.
35   OFCOM Convergence Think Tank 1, Ed Richards, Chief Executive, OFCOM, 07/02/2008.

The adoption of the AVMS Directive followed a period of two years in which the European Commission reviewed the Television Without Frontiers (TVWF) Directive[36] following its last revision in 1997, in light of the advent of convergence in the media world. The AVMS Directive provides new rules on forms of commercial communications which are designed to enable flexibility in the financing of the industry whilst ensuring consumer protection. The new Directive aims to promote a level playing field in Europe between traditional and new media services, creating legal certainty and increasing competition, getting the best for consumers. The European Commission is keen to promote media pluralism and not to stifle technological innovation.

Scope

The AVMS Directive amends the existing TVWF legislation and expands its scope, which was limited to covering the transmission of television programmes intended for the public by wire or over the air (to cover satellite transmissions), to apply to all forms of audiovisual media service. Specifically, it extends the existing framework to cover on-demand services. It classifies audiovisual media services as being either linear (called 'television broadcasting') or non-linear (called 'on-demand' services). The AVMS Directive has a two tier regulatory regime and differentiates between these two types of service. Despite the expansion in the application of the AVMS Directive to a wider source of media, not all audiovisual media services will be covered by the new rules. In its definition of an audiovisual media service, the Directive sets out a number of requirements which a service must satisfy in order to be covered by the rules.

- The media service provider ('the natural or legal person who has editorial responsibility for the choice of the audiovisual content of the audiovisual media service and determines the manner in which it is organised') operating the audiovisual media service must be based in the EU. Editorial responsibility is defined as:

  'the exercise of effective control both over the selection of the programmes and over the organisation either in a chronological schedule, in the case of television broadcasts, or in a catalogue, in the case of on-demand audiovisual media services.'

- The audiovisual media service must consist of programmes. Programmes are defined as:

  'a set of moving images with or without sound constituting an individual item within a schedule or a catalogue established by a media service provider and whose form and content is comparable to the form and content of television broadcasting';

- The programmes must have as their principal purpose 'to inform, entertain or educate' the general public;

- This must occur through the medium of 'electronic communications networks', as defined in the Framework Directive (2002/21/EC) as:

  'transmission systems and, where applicable, switching or routing equipment and other resources which permit the conveyance of signals by wire, by radio,

---

36   89/552/EEC.

> *by optical or by other electromagnetic means, including satellite networks, fixed (circuit- and packet-switched, including Internet) and mobile terrestrial networks, electricity cable systems, to the extent that they are used for the purpose of transmitting signals, networks used for radio and television broadcasting, and cable television networks, irrespective of the type of information conveyed'.*

- As the Department of Culture, Media and Sport commented in its consultation on the AVMS Directive, the reference to 'electronic communications networks' and the application of the above definition makes the neutral stance the Directive takes towards technology clear. Arguably, the original TVWF Directive was ambiguous in its scope, covering 'services provided by wire or over the air'[37];

- On-demand services (non-linear services) must be 'television-like'. They must:

  > *compete for the same audience as television broadcasts and the nature and the means of access to the service would lead the user reasonably to expect regulatory protection'; and*

- The audiovisual media service must be mass media, that is, it must be 'intended for reception by, and which could have a clear impact on a significant proportion of the general public'.

Another important definition under the Directive is that of an 'audiovisual commercial communication', which are defined as:

> *'images with or without sound which are designed to promote directly or indirectly, the goods, services or image of a natural or legal entity pursuing an economic activity. Such images accompany or are included in a programme in return for payment or for similar consideration or for self-promotional purposes.'*

Examples, which the Directive gives, are of television advertising, sponsorship, teleshopping and product placement.

There has been criticism of the scope of the Directive. Arguably the definition of audiovisual media service does not go far enough, meaning that the Directive is not 'future proof'. With the ever increasing popularity of shows broadcast solely on sites such as YouTube it appears an increasing number of services will not be covered by the new Directive. These issues will be discussed later.

The exact scope of editorial responsibility is as of yet unsure. The AVMS Directive gives Member States the power to shape this definition and notably the definition of 'effective control'. This will prove to be key since as the definition stands it is not clear, in certain scenarios, who has the responsibility for complying with the rules the AVMS Directive lays down. For example, many services may have more than one entity involved in the distribution of the content, such as where there is a content aggregator. There has been some doubt as to whether content aggregators would be regulated. A further complication is that no change is made to the exemptions from liability which are established in the E-commerce directive[38] where a service provider can be exempted from any responsibility for compliance if they act as a

---

37 DCMS Consultation on Implementing the EU AVMS Directive, Part 2 Scope: The Definition of Audiovisual Media Service.
38 Directive 2000/31/EC, Articles 12 and 13.

'mere conduit' or if they are merely 'caching' the data and are 'in no way involved with the information transmitted'. With the increase in service providers distributing audiovisual content through their websites, the boundary as to whether they are a 'mere conduit' or whether they have editorial responsibility for the content is becoming increasingly blurred.

Application for Satellite

The AVMS Directive also provides that broadcasters who either use a satellite uplink from a Member State or use satellite capacity appertaining to that jurisdiction will be considered to be 'established' in that Member State for the purposes of the AVMS Directive. Such broadcasters are required to comply with the requirements of the AVMS Directive as applied in the relevant Member State.

Distinction Between Linear and Non-linear

The AVMS Directive divides programmes into two categories, television broadcast (or linear) and on-demand (or non-linear). It defines a television broadcast as a linear audiovisual media service which is 'provided by a media service provider for simultaneous viewing of programmes on the basis of a programme schedule'. Viewers have no choice of what they watch and in what order, the schedule is dictated by the media service provider and content is 'pushed' to them. The AVMS Directive identifies analogue and digital television, live streaming, webcasting, and near-video-on-demand as forms of television broadcasting. The delivery of linear content is not restricted to traditional forms of delivery such as the television. This category would also cover the broadcasting of a traditional television channel through a mobile phone streaming service which seems likely to become popular in the future.

Non-linear programmes are on-demand audiovisual media services provided:

> 'for the viewing of programmes at the moment chosen by the user and at his individual request on the basis of a catalogue selected by the media service provider.'

The user therefore decides when to watch programmes and 'pulls' content from the media service provider's database. Such non-linear programming has recently become very popular in Europe through video-on-demand services.

Two-tier Approach

The European Commission made clear in its proposals prior to the implementation of the AVMS Directive that it believed that non-linear content should receive lighter regulation than linear content, to reflect the differences in user choice and control and the likely impact on society[39]. This approach differentiates regulation on the basis that services 'pushed' to a consumer should be more highly regulated than those where the consumer 'pulls' content from a provider and exercises a level of

---

39 Commission Proposal for a Modernisation of the Television without Frontiers Directive: Frequently Asked Questions MEMO/06/208.

personal choice. Critics have suggested that the line between this 'push' and 'pull' distinction is not as clear as the Directive envisages. The argument is that there is very little difference between a consumer recording a television broadcast on their PVR (personal video recorder eg Sky+/V+) and watching it at a later date and the consumer accessing the same programme on one of the growing number of 'video-on-demand' services such as the BBC's iPlayer. The former would, under the AVMS Directive, be subject to a greater level of regulation, as an example of linear content, whilst the latter would be the subject of less stringent regulations, as an example of non-linear content.

Some rules apply universally to all audiovisual media services, both linear and non-linear, and there are also an additional set of rules which apply solely to linear audiovisual media services.

Rules Applying to Both Linear and Non-Linear Services

Country of Origin principle: The country of origin principle which applied to television broadcasters under the TVWF—that a service provider only comes under the jurisdiction of the regulator in its own Member State even if the service is made available in other Member States—will be extended to cover all audiovisual media services. The European Commission believes that this principle has enabled Europe's broadcasting industry to flourish since 1989 and it hopes that going forward the same will happen to non-linear content. The practical impact of the rule is that audiovisual media service providers will be able to offer content complying with the laws of their Member State for broadcast in other Member States without having to modify the content in order to comply with local laws in each country. The aim of this is to facilitate a level playing field and to increase diversity and choice for the consumer. The AVMS Directive further develops the principle established in the TVWF Directive that a Member State may apply stricter rules on media service providers based in its jurisdiction than those set out by the Directive. These are now further qualified by restrictions aimed at preventing 'jurisdiction shopping'. For example, where a Member State has stricter rules governing media service providers within its jurisdiction and it has determined that a broadcaster under the jurisdiction of another Member State has been providing a television broadcast that is 'wholly or mostly directed towards its territory', it is able to contact the other Member State and attempt to resolve the problem through a series of steps designed to operate in the general public interest, set out in part of the AVMS Directive which amends Article 3 of the TVWF Directive.

Freedom of Reception: The AVMS Directive extends the application of the principle of freedom of reception established under the TVWF Directive to all audiovisual media services. The AVMS Directive extends the principle to cover on-demand (non-linear services). Member States can restrict access in respect of a given service which prejudices or presents a serious risk of prejudicing a number of public policy related reasons, for example the prevention criminal offences, protection of minors, human dignity and incitement to hatred on grounds of sex, race, religion or nationality or for reasons of national security. The measures taken to restrict freedom of reception on these grounds must be proportionate and as with television broadcasting there needs to be communication with the offending broadcaster, the affected Member State and the European Commission before any restriction may be implemented. There is a limited right to restrict such freedom in case of urgency.

Content standards, Protection of Minors: Provisions of the TVWF Directive protecting minors are retained by the AVMS Directive. In addition it now makes it clear that there should be a balance between the measures taken to protect the physical, mental and moral development of minors and human dignity and the fundamental right to freedom of expression. Suggestions of such protection methods include the use of PIN codes, filtering systems or labelling. On-demand services which contain content which is not appropriate for minors must have measures to prevent them from viewing or hearing such content.

Disability access: The AVMS Directive introduces a requirement that 'services are gradually made available to people with hearing or visual disability'. This reflects the growing importance of providing access for people with disabilities as a core duty of broadcasters. The AVMS Directive does not provide any further clarification on timescale of 'gradually' or the steps required to make the service available. Member States will need to interpret this obligation and provide more detailed obligations when implementing it.

Product Placement: Rules regarding product placement have been liberalised by the AVMS Directive as a result of the European Commission's drive to make EU audiovisual services more competitive. This is partly in light of the increasing share of advertising revenue that this type of advertising has in the USA. The European Commission has stated that in the USA:

> 'product placement accounts for 1.7 per cent of the total advertising revenues of free-to-air broadcasters... and had grown by an average of 21 per cent per year between 1999 and 2004.'[40]

In a further European Commission release[41], the European Commission claimed that:

> 'recent figures from countries that permit product placement suggest that clear rules should help the European audiovisual industry to become more competitive, especially compared with the USA.'

They further claim that:

> 'an independent study published in March 2007 found that global paid product placement grew 37 per cent in 2006 and is forecast to grow 30 per cent in 2007. Television placements remain the dominant choice of brand marketers, accounting for 71 per cent of global spending.'

The AVMS Directive aims to level the playing field in the EU with regards to product placement since certain countries currently allow it and others prohibit it. The new rules aim to improve certainty for media service providers and provide protection for consumers.

The default position under the AVMS Directive is that product placement is prohibited. However, it is admissible, unless Member States decide otherwise:

---

40 The Commission Proposal for a Modernisation of the Television Without Frontiers Directive: Frequently Asked Questions, MEM/06/208, quoting figures from the 'Comparative study on the impact of control measures on the television advertising markets in European Union Member States and certain other countries', pp 60–61.

41 Presenting the Audiovisual Media Services without frontiers Directive: Frequently asked questions, MEMO/07/206 of 24 May 2007.

(i) in cinematographic works, films, television series, sports programmes and entertainment programmes (but not children's programmes); or (ii) in all cases so long as there is not payment for the product's 'placement' (but instead it takes the form of production props without charge). Member States are able to choose the extent which they permit product placement within these boundaries, for example by allowing product placement only in programmes which have been produced outside the European Union or only for non-linear services.

The European Commission was keen to ensure that product placement is not excessively intrusive. Products must not have undue prominence. Content and scheduling of programmes must not be influenced so as to affect editorial responsibility and independence of the media service provider and the product placement must not directly encourage purchase or rental of goods or services, with a specific prohibition on promotional references to the goods or services. One of the key requirements is that viewers are clearly informed of the existence of the product placement at the beginning and end of the programme as well as after every advert break so that they can distinguish between the editorial content and the product placement. This final requirement may be waived by Member States if the programme in question has been produced independently of the media service provider which must also have no affiliation to the production company.

An absolute prohibition is placed on the product placement of tobacco products, cigarettes and prescription only medicinal products or treatments. Even product placement by undertakings whose main activity is the manufacture or sale of cigarette or tobacco products is prohibited.

The Department of Culture, Media and Sport in the UK has decided that it will retain the current position on product placement. As a result, product placement will continue to be banned in linear programmes made for or by a UK television broadcaster although the UK government will review this again in 2011/2012. Under the OFCOM Broadcasting Code, UK television broadcasters are able to show cinema films and non-UK television programmes which contain product placement and this will continue to be the case. Product placement will also continue to be permitted for non-linear programmes. This distinction between linear and non-linear content could mean that product placement could be used in the on-demand version of a programme while being prohibited in the television broadcast.

Access to Information: The AVMS Directive introduces an obligation for audiovisual media service providers, whether linear or non-linear, to provide certain information to viewers. This includes the service provider's name and address, their email or web address and details of the competent regulatory or supervisory body, where applicable. The justification for the provision of this information is that these services have an impact on the way people form their opinions and an element of accountability is required. Member States have freedom in deciding practical details in this area.

Quotas: The rules regarding quotas on European works set out in the TWF Directive remain with respect to linear programming following the AVMS Directive's implementation. There is now also a similar requirement with respect to non-linear programming. This regulates the amount of content that is classified under the still very broad term of European Works and either takes the form of the level of financial contribution a service makes to the production or rights acquisition of such works or it takes the form of a share of the services catalogue that is dedicated to showing such works. Reports must be sent to the European Commission every four years detailing the implementation of this initiative. It is difficult to see what effect this will have for non-linear programmes. It is expected that this would apply to the per centage of European Works in the catalogue rather than the programmes viewed, as viewers have control over which programmes they select to watch.

Advertisements: The AVMS Directive relaxes some of the regulations on advertising and passes some of the decision making to the broadcasters themselves. Television advertising, sponsorship, teleshopping and product placement are all included under the umbrella definition of 'audiovisual commercial communication'. In its proposals leading up to the AVMS Directive, the European Commission suggested that market forces would provide a form of regulation for the industry. In an increasingly competitive audiovisual media industry consumers would in effect vote with their remote controls if broadcasters abused the new advertising flexibility offered to them. Restrictions applying to advertising are characterised as either qualitative or quantitative.

The **qualitative restrictions** apply to both linear and non-linear services. A clear line must be drawn between television advertising or teleshopping and any editorial content. The prohibition on advertisements prejudicing respect for human dignity, or on including or promoting discrimination established under the TVWF Directive are extended to cover new forms of discrimination, such as age discrimination. Rules regarding health and safety and the environment are retained. All advertisements for cigarettes and tobacco products are still prohibited and this ban is extended to non-linear services. Alcohol advertising still carries restrictions. The AVMS Directive also maintains the blanket prohibition on advertising for prescription-only medical products and treatment.

Rules under the TVWF Directive requiring advertising and programmes to be kept separate cannot be strictly applied under the AVMS Directive as a result of the introduction of product placement. But the AVMS Directive does require that advertising be readily distinguishable from editorial content.

Following steadily increasing concerns regarding healthy eating, the AVMS Directive encourages Member States and the European Commission to work with media service providers to develop codes of conduct regarding the promotion of unhealthy 'junk food' linked with children's programming. Following research carried out by OFCOM in 2006, scheduled restrictions on the advertising of high fat, salt and sugar (HFSS) food and drink to children came into effect in February 2007 as part of a phased set of restrictions imposed on the advertising industry. Since January 2008, there has been a total ban on HFSS advertisements in programmes aimed at children aged 4–15, or those attracting disproportionately high child audiences, and a requirement that children's channels scale back the advertising of such products to 50 per cent of 2005 levels. The final step of the three phases put in place a total ban on such advertising on children's channels from January 2009. OFCOM has produced interim data to show that the new rules appear to be having the desired effect on reducing the quantity of HFSS food and drink advertising that children are exposed to on television. A report[42] has been put together by OFCOM containing a full review of how the new restrictions are working, covering data from January to June 2008, and thus reflecting the impact of the second phase of restrictions.

---

42   'Changes to the nature and balance of television food advertising to children' http://www.ofcom. org.uk/research/tv/reports/hfssdec08/.

Content Regulation

Sponsorship rules under the AVMS Directive extend those established under TVWF to non-linear services. The reference to independent scheduling is retained in relation to linear broadcasts alone. In addition, Member States may now decide whether to prohibit sponsorship logos being shown during children's programmes, documentaries and religious programmes.

Surreptitious advertising is also prohibited under AVMS, as it was under TVWF.

Rules Only Applying to Linear Services

Quantitative advertising restrictions only apply to linear audiovisual media services—they have not been extended to cover non-linear services. The European Commission is keen to maximise the competitive nature of EU based audiovisual services and it recognises that advertising is an important source of revenue for the industry, enabling it to invest in future works and develop new technologies, thereby competing effectively with companies across the world. This outlook has lead to a relaxation in the quantitative restrictions surrounding audiovisual advertisements.

As part of this added flexibility, media service providers will now be able to choose the best moment to insert advertisements and the previous rule requiring a minimum of 20 minutes between advert breaks has been removed. The AVMS Directive marks a shift in emphasis from regulators to broadcasters when deciding how best to retain the integrity of the programming. Children's programming is not afforded this increased flexibility and such programmes may be interrupted once in a period of 30 minutes only if the programme lasts longer than 30 minutes. Films and news programmes have a relaxed regime, and can now be interrupted every 30 minutes, as opposed to the previous 45 minute limit. However, despite this apparent increased flexibility, the overall restriction of 12 minutes (20 per cent) of advertising in every hour of broadcasting is still applicable.

High Interest Events:- The rules regarding events of major importance originating under the TVWF Directive still apply. In addition, the AVMS Directive provides that those media service providers with exclusive broadcast rights to an event of high interest to the public must allow other broadcasters to use short extracts for general news programmes (and not those programmes with entertainment as a purpose). These extracts should not exceed 90 seconds and whilst in theory the right of access to such clips is EU-wide, it should only be granted cross-border where necessary and only when the broadcaster seeking access has exhausted the options in its Member State. It should be possible for media service providers to provide their 'live' broadcasts containing these short clips as an on-demand product following the live event without the need to edit the programmes and remove the short clips.

Co- and Self-Regulation: For the first time, the AVMS Directive encourages use of self-regulation (the adoption of common guidelines by groups such as social partners and economic operators) and co-regulation. The Department of Culture, Media and Sport explains co-regulation in this context to be:

'an arrangement under which the v-o-d industry regulates itself, devising and introducing Codes or other mechanisms to ensure that EU content standards are met …. But with a power for the public authorities to intervene in the event

of a serious and sustained failure to meet the content standards of the (AVMS) Directive.'[43]

This is an important principle in an industry which is constantly evolving and where traditional forms of regulation may not be able to keep up to speed with the pace of technological advancements. This is not, however a substitute for the obligations of the national regulator.

Evaluation

The aim of the AVMS Directive was to update the law in the area of content regulation to reflect the fast paced technological advances that have occurred and are continuing to occur since the last revision of the TVWF Directive. It was intended to bring greater legal certainty as to the boundaries of what is permitted, particularly in relation to new media services as well as updating some of the legal burden on so called traditional media services. Although the AVMS Directive extends the regulatory framework to new media services, there are some uncertainties as to how it will apply in practice to these services. The AVMS Directive is an extension of the TVWF framework and as such it was principally developed to regulate traditional broadcasting. When this framework is applied to some of the new, emerging services, particularly on the internet and mobile, it may become apparent that the regulatory structure does not entirely reflect the business and technological model of these services and difficulties of interpretation and application may arise.

Practical Issue—the AVMS directive and mobile television

Although the AVMS Directive is intended to be platform neutral, it may have unintended effects which means that it does not apply equally to all platforms. An example of this is mobile TV. Mobile TV is a growing service which although it may be linear (ie broadcast on a schedule), is very different in structure and economic model from traditional broadcast television. Some of the requirements imposed on mobile TV where it is a linear service do not fit well with the nature of the service. The quantative restrictions on advertising, only allowing 12 minutes of advertising per hour or restricting adverts to every 20 to 30 minutes do not work well for mobile TV, which viewers tend to watch for only a few minutes at a time. Such advertising restrictions could mean many viewers miss the adverts altogether. This could make the mobile TV service uneconomical if the advertisers reduced the amount they would pay as a result. It is also difficult to see how the accessibility rules such as requiring subtitles would work in practice. Mobile screens are small and subtitles are not very practicable. Another area of concern is short reporting of high profile events. If a mobile TV company has exclusive rights to an event, the AVMS directive will require it has to grant short reporting rights to other broadcasters for clips of up to 90 seconds. As the mobile TV service's own coverage will be short extracts of only a few minutes, granting such short reporting rights could undermine its exclusivity. Finally on this point, it makes little sense to regulate on-demand and scheduled mobile television services differently—consumers see very little distinction because either way they will be choosing what to watch (for a few minutes only) from a menu.

---

43 DCMS Consultation on Implementing the EU AVMS Directive Part 3 Section Regulation of on-demand Audiovisual Media Services p. 37.

> The concern for all these points is that not only will there be difficulty applying AVMS Directive to such mobile TV services but the requirements of the AVMS Directive may actually harm the fledgling economic model of these services.

The requirement for audiovisual media services to be 'television-like' is vague. They must be comparable to television broadcasting; this is a fairly dynamic concept, which the Directive acknowledges in the recitals. The question is how television broadcasting is going to be defined, whether as television as we currently know it or television as it evolves in the future.

The Directive prescribes that the audiovisual media services must be 'mass media' in that they are intended for reception by and could have a clear impact on a significant proportion of the general public. This is a particularly different concept to quantify going forward. Firstly, who is the general public? Is this the general on-line public anywhere in the world? Is this the EU general public? Or is this the UK general public? Furthermore, who is going to measure it? Who is going to decide what a clear impact is and what has the potential to have a clear impact? The traditional broadcast channels have seen declining viewing numbers since the rise in the number of new media outlets; how large an audience must be to constitute a 'significant proportion of the general public' is not clear.

User generated content (UGC) is expressly excluded by the Directive since it is non-commercial and the content is generated by private individuals for sharing and exchange within communities. The result of this is that much of the content posted on sites such as YouTube and Bebo will not be covered. This approach does not appear to be 'future proof'; UGC has increasingly large viewer numbers, especially amongst minors and young people. Avoiding regulation of such content would provide a back door by which children could be exposed to the very content which the Directive seeks to protect them from. As an example, a series on the social networking site Bebo called 'Kate Modern' comprised 155 episodes each lasting under four minutes. The show drew a total audience of more than 35 million episodes viewed, which averaged as 1.5 million per week. This content would be unlikely to be regulated by the AVMS Directive since it is likely to be deemed not to be sufficiently 'television-like', yet a small network television show drawing fewer viewers would be covered.

The Directive also states that for an audiovisual media service to be 'television-like' it must compete for the same audience as television broadcasts and the nature and means of access to the service will leave the user to reasonably expect regulatory protection within the scope of the Directive. It appears that whilst in the past new media has competed with television services for the same audiences, in the future it is likely that television services will be competing with new media for audiences. The wording of this recital appears to be a little back to front and does not reflect forward thinking. With respect to a user's reasonable expectation of regulatory protection, who decides what user expectations are going to be? People of all ages and demographics have differing expectations, so how are these going to be evaluated?

All these challenges will mean that national regulatory authorities will have a tough task in interpreting the Directive to implement it into their national law. It is important to remember that this Directive is being implemented in 27 different

Member States that all operate in very different regulatory environments and from different starting points. The risk here is that the AVMS Directive does not fulfil its aim of harmonisation.

Following consultation, the Department for Culture, Media and Sport released a ministerial statement providing an outline of how the AVMS Directive will be implemented into English law[44]. This provides the general principles for implementation but many of the details still need to be clarified prior to implementation in December 2009. At the time of writing, the intention is to implement the AVMS Directive as follows:

Regulation of linear services: These will continue to be regulated by OFCOM and the ASA in the same way as before, with very few changes. The most significant change of linear services is likely to be a relaxation on the scheduling of advertising.

Regulation of non-linear services: These will be regulated by a new co-regulatory body to be established by OFCOM before the end of 2009. It is not clear whether ATVOD will provide the basis for the new body but ATVOD is likely to play a significant role in its establishment. The Department for Culture, Media and Sport intends to adopt the Directive's definitions on the scope of non-linear services regulated and leave it to the new co-regulatory body to issue guidance on scope and interpretation. However, the Department for Culture, Media and Sport did indicate that these definitions should be given a narrow scope and it has expressly excluded pure content aggregators. As a result, the uncertainties as to which non-linear services will be regulated are likely to continue until the new co-regulatory body issues detailed guidance. The ASA will regulate advertising on or around non-linear programming, applying the CAP Code.

Regulation of foreign channels: OFCOM will regulate non-UK satellite channels up-linking from the UK for the first time, requiring them to obtain a licence and comply with the UK regulation on content and advertising. The only exception will be where the non-UK satellite channels are already regulated by another EU country. The UK-based up-linkers for these channels will not be regulated. The up-linkers will be required to comply with an order from OFCOM to stop uplinking for a certain channel but will not be required to check the channel's compliance with its licence or to monitor its content.

The Department for Culture, Media and Sport is working with OFCOM and the ASA to draw up the legislation to implement the AVMS Directive and the guidance surrounding it. It is expected that the implementing legislation will be laid before the UK Parliament in the summer and will be passed by the end of 2009. There will be special provisions for the BBC and SC4. It will only be after the legislation is finalised and the core guidance published that it will be clear exactly how the UK has implemented the AVMS Directive.

---

44   The statement can be found at http://www.culture.gov.uk/reference_library/minister_speeches/5932.aspx/.

Chapter 7

Wireless Telegraphy

Claire Wright and Thomas Butcher   ALLEN & OVERY
Allen & Overy LLP

7.1 Introduction to Wireless Telegraphy

Radio spectrum (a term which is generally understood to mean the electromagnetic frequency range from 3 kHz to 300 GHz) is a finite, and increasingly valuable, resource. As the medium through which wireless communications are enabled, it has a critical role to play in our daily lives. It has a multitude of uses, including radio and television broadcasting, defence, and fixed, mobile and satellite telecommunications.

The importance of radio spectrum as an economic asset and the fact that demand for it is increasing all the time means that access to and use of it must be appropriately planned and managed. Failure to do so inevitably results in multiple users transmitting on the same frequencies at the same time causing interference between users to the point that services that depend on radio spectrum are significantly impaired and, at worst, rendered unusable.

This chapter is intended to provide an overview of the regulatory framework governing the management and use of radio spectrum in the United Kingdom.

7.2 The Regulatory Regime for Radio Spectrum

The primary sources of radio spectrum regulation in the United Kingdom are national laws: the Wireless Telegraphy Act 2006 (the WTA 2006) and the Communications Act 2003 (the Communications Act), each of which has a raft of associated regulation. These Acts were enacted, and should be viewed in the context of, relevant global and European frameworks.

7.2.1 The Wireless Telegraphy Act 2006

The WTA 2006, which came into force on 8 February 2007, consolidated existing spectrum regulation, which had developed in a piecemeal fashion over a number of years[1]. It sets out OFCOM 's functions, duties and enforcement powers in relation to radio spectrum and establishes the regime for licensing and management of that spectrum. The WTA 2006 is discussed in more detail in section 7.3.

7.2.2 The Communications Act 2003

The Communications Act, which implements a series of European Directives[2], sets out the powers and duties of OFCOM. This includes those applicable to OFCOM's performance of functions under what is now the WTA 2006. In addition to its general duties in relation to communications matters[3], OFCOM has spectrum-specific powers and duties—described in further detail in section 7.3. OFCOM's performance of its functions is subject to any directions given by the Secretary of State[4].

Until 2007, the Communications Act set out OFCOM 's functions and enforcement powers in relation to spectrum management and use. The relevant provisions were repealed and superseded by the WTA 2006.

7.2.3 Responsibility for the Regulation of Radio Spectrum in the United Kingdom

Part 1 of the WTA 2006 identifies the general functions of OFCOM in relation to the management and use of the radio spectrum. However, OFCOM is only tasked with managing radio spectrum allocated for civil use in the United Kingdom. A number of public bodies (including the Ministry of Defence and the Civil Aviation Authority) are responsible for managing the remainder of it.

7.2.4 European Radio Spectrum Policy

The European Commission (the Commission) has no direct responsibility for spectrum management, but oversees a number of measures designed to co-ordinate policy approaches and work towards harmonised conditions for the availability and efficient use of radio spectrum.

---

1 The principal pieces of primary legislation being the Wireless Telegraphy Acts 1949, 1967 and 1998, the Marine etc Broadcasting (Offences) Act 1967, Part 6 of the Telecommunications Act 1984 and certain provisions of the Communications Act.
2 Directive 2002/21/EC on a common regulatory framework for electronic communications networks and services (the **Framework Directive**), Directive 2002/20/EC on the authorization of electronic communications networks and services (the **Authorisation Directive**), Directive 2002/19/EC on access to, and interconnection of, electronic communications networks and associated facilities (the **Access Directive**) and Directive 2002/22/EC on universal service and users' rights relating to electronic communications networks and services (the **Universal Service Directive**). As at the time of writing proposed reforms of those Directives are the subject of discussion between the European Commission and the European Parliament.
3 These are duties to: (a) further the interests of citizens in relation to communications matters; and (b) to further the interests of consumers in relevant markets, where appropriate by promoting competition: Communications Act 2003, section 3(1), and to promote the requirements set out in the Framework Directive: Communications Act 1993, section 4.
4 Communications Act 1993, section 5.

At the time of writing, the mechanisms for achieving these regulatory aims are established in the 2002 Radio Spectrum Decision[5].

The Radio Spectrum Decision establishes the Radio Spectrum Committee[6]. Composed of Member State representatives and chaired by the Commission, the Committee assists the Commission in the development of decisions on technical implementation measures and the propagation of information about spectrum use.

The work of the Radio Spectrum Committee is complemented by that of the Radio Spectrum Policy Group (RSPG)[7]—set up in 2002 and comprised of subject matter experts from Member States. The RSPG advises the Commission on spectrum-related issues such as availability, harmonisation, allocation, availability and use of radio spectrum, methods for granting rights of use, refarming, relocation, valuation and efficient use and protection of human health.

The Radio Spectrum Decision empowers the Commission to issue mandates to CEPT—the European Conference of Postal and Telecommunications Administrations—for detailed development of implementing measures[8]. Once developed, implementing measures are given effect through Commission Decisions, which are binding on Member States, including the United Kingdom.[9] These Commission Decisions form the basis for the pan-European harmonisation of spectrum usage at an applications level.

In addition, the United Kingdom is required to implement relevant Directives, the principal of which is the GSM Directive (Directive 87/372/EEC). The GSM Directive reserved certain frequency bands for the co-ordinated introduction of GSM mobile services on a pan-European basis. As a consequence of the development of digital technologies capable of co-existing with GSM in the relevant bands there is, at the time of writing, a formal proposal to amend the GSM Directive to allow the use of the 900 MHz band for the provision of various pan-European services including UMTS (3G) services. This is an example of a process referred to as 'spectrum refarming' in which radio spectrum allocated for one type of use is cleared and reassigned to (typically) higher value applications.

The Commission also plays a role in monitoring relevant radio spectrum developments outside the European Union and in international organisations. It participates in the World Radiocommunications Conferences as a non-voting delegate.

7.2.5 EU Telecoms Reform

In November 2007, the Commission proposed a package of reforms to the EU's regulatory framework for electronic communications networks and services aimed

5   Decision 676/2002/EC of the European Parliament and of the Council of 7 March 2002 on a regulatory framework for radio spectrum policy in the European Community.
6   Radio Spectrum Decision, Article 3(1).
7   The RSPG has a dedicated website at http://rspg.ec.europa.eu/.
8   The body within CEPT responsible for spectrum planning is the European Communications Committee (ECC).
9   A list of Decisions in force is available on the Commission website, at http://ec.europa.eu/.
     information_society/policy/radio_spectrum/ref_documents/index_en.htm#.

at creating a single EU telecoms market, increasing competition, improving rights for consumers and businesses and promoting the deployment of cross border services.

Specific proposed reforms impacting radio spectrum policy include:

- amendments to Article 9 of the Framework Directive to make technology neutrality a binding principle and to introduce the principle of service neutrality (so that a frequency band which may only currently be used to provide a specific service can be used for any service adopting any technology);

- a mechanism to allow spectrum to be traded in certain designated EU wide frequency bands;

- regulatory provisions to encourage licence-free spectrum use and to reinforce the coordination of conditions for spectrum authorisations; and

- the submission by the Commission of a multi-annual EU radio spectrum policy programme to be jointly adopted by the European Parliament and Council.

The European Parliament had suggested that a new advisory body for radio spectrum be created although this proposal has not been retained by the Commission on the basis that it would potentially duplicate the work undertaken by the existing RSPG. At the time of writing, the future of the telecoms Framework Review is uncertain. See chapter 1 for more details.

7.2.6 International Organisations

At an international level, radio spectrum falls within the remit of ITU-R, the part of the International Telecommunications Union (ITU) with responsibility for the radiocommunication sector. The ITU is a United Nations agency whose membership comprises nation states, public and private sector companies and international and regional telecommunications entities[10]. The ITU is tasked with allocating broad category uses to specific frequency blocks (typically referred to as frequency bands).

ITU-R's stated mission is to:

> *'ensure the rational, equitable, efficient and economical use of the radio-frequency spectrum by all radiocommunication services, including those using satellite orbits, and to carry out studies and approve Recommendations on radiocommunication matters'.*

The principal output of ITU-R deliberations is the Radio Regulations (the most recent edition of which was issued in 2008), which have treaty status[11]. ITU-R reviews and updates the contents of the Radio Regulations through World Radiocommunication Conferences, held every two to three years. No sanctions are provided for breach of the Radio Regulations.

The Radio Regulations include a frequency table (known as the International Table of Frequency Allocations) which identifies spectrum allocations. Typically allocations

---

10 The ITU's history, constitution and mission are described in more detail on its website, www.itu.int.
11 The Radio Regulations are available for purchase on the ITU's website, at http://www.itu.int/publ/R-REG-RR/en.

are not technology or application specific (so, for example, allocations are to broadcasting generally rather than to television or radio broadcasting). In addition to the Table of Frequency Allocations, the Radio Regulations contain provisions intended to promote transmission of radio services and minimise interference between countries. Crucially though, nothing in the Radio Regulations prevents an individual country from regulating spectrum as it wishes, provided this does not have anything other than a minor impact on other countries and it is itself prepared to accept the risk of interference.

7.3 Overview of the Wireless Telegraphy Act 2006

Radio frequency spectrum use in the United Kingdom, Channel Islands and the Isle of Man is governed by the WTA 2006, which specifies that any person using or installing radio equipment must be licensed by OFCOM. As noted in section 7.2, the WTA 2006 consolidates all legislation relating to the management of radio spectrum into a single Act. Therefore it replaced the Wireless Telegraphy Acts 1949, 1967 and 1998, the Marine etc Broadcasting (Offences) Act 1967, Part 6 of the Telecommunications Act 1984 and certain provisions of the Communications Act 2003 (Communications Act), making very few substantive changes to the previous legislation governing spectrum management and usage. Wireless telegraphy licences issued and regulations made before 8 February 2007 have continued in force as if made under the WTA 2006.

7.3.1 Radio Spectrum Functions of OFCOM

OFCOM's general powers with regard to spectrum management are stated in sections 1 and 3 of the WTA 2006. It is a function of OFCOM to provide whatever advice and services it considers appropriate for facilitating or managing the use of radio spectrum for wireless telegraphy[12] and grant spectrum to any person who holds a wireless telegraphy licence or a grant of recognised spectrum access or to any other person if, in each case, OFCOM is of the opinion that the making of the grant is likely to promote the efficient use or management of spectrum in the United Kingdom[13].

Other responsibilities and functions of OFCOM include:

■ a requirement to publish the United Kingdom Plan for Frequency Authorisation (the Frequency Authorisation Plan) which sets out the frequencies that have been allocated for particular wireless telegraphy purposes, frequencies available for assignment and the purposes for which the different frequencies have been allocated[14];

---

12 Defined as the transmission of electromagnetic energy over a frequency not exceeding 3,000 GHz and over paths that are not provided by a material substance constructed or arranged for that purpose in order to convey messages, sound or visual images; operate or control machinery or apparatus; determine position, bearing or distance; or gain information as to the presence, absence, position or motion of an object or class of objects, section 116.

13 Wireless Telegraphy Act 2006, section 1(5)–(6).

14 Wireless Telegraphy Act 2006, section 2. The Frequency Authorisation Plan is published by OFCOM on its website, at http://spectruminfo.ofcom.org.uk/spectrumInfo/ukpfa.

- the provision of an interference service consisting of the giving of advice and assistance to persons complaining of interference with wireless telegraphy[15]; and

- carrying out research to ascertain the current and likely future demand for use of radio spectrum for wireless telegraphy, the effects of any such use and any other relevant matters that OFCOM deems relevant[16].

Section 3 of the WTA 2006 sets out the duties OFCOM must comply with when carrying out its functions. In undertaking its spectrum management role, OFCOM must consider the availability of, and present and future demand for, spectrum. It must also have regard to the desirability of promoting the efficient management and use of spectrum (which, at its simplest level, means preventing interference between competing users of the radio spectrum), the economic and other benefits that can arise from the use of spectrum, the development of innovative services and competition in the provision of electronic communications services[17]. Where it appears to OFCOM that there is a conflict between its duties under section 3 of the WTA 2006, it is required to resolve the conflict in the manner it thinks is best in the circumstances[18]. If it appears to OFCOM that there is a conflict between a duty under section 3 of the WTA 2006 and one or more of its overarching duties under sections 3 to 6 of the Communications Act, its duties under the Communications Act must prevail[19].

OFCOM's management of the radio spectrum is constrained by any directions given by the Secretary of State. Indeed the Secretary of State can require OFCOM to reserve specific frequencies for particular uses or users[20]. The author is not aware that this power has ever been exercised.

7.3.2 Requirements for a Wireless Telegraphy Licence

The WTA 2006 states that it is unlawful for any person to establish or use a wireless telegraphy station[21], or install or use wireless telegraphy apparatus[22], without being licensed by OFCOM[23]. A wireless telegraphy licence can be granted in relation to a particular station or apparatus or in relation to any station or apparatus falling within a description specified in the licence[24].

7.3.3 Licence Exemptions

Section 8 adds a number of exceptions to the general rule requiring a wireless telegraphy licence. First, the use or installation of television receivers for receiving television programmes does not require a licence[25]. Secondly, OFCOM has the power by regulations to grant either an absolute or conditional exemption for

---

15 Wireless Telegraphy Act 2006, section 4.
16 Wireless Telegraphy Act 2006, section 1(4).
17 Wireless Telegraphy Act 2006, section 3(1)–(2).
18 Wireless Telegraphy Act 2006, section 3(6).
19 Wireless Telegraphy Act 2006, section 3(5).
20 Wireless Telegraphy Act 2006, section 5(1)–(2).
21 A station for emitting or receiving wireless telegraphy, section 117(2).
22 Apparatus for emitting or receiving wireless telegraphy, section 117(1).
23 Wireless Telegraphy Act 2006, section 8(1).
24 Wireless Telegraphy Act 2006, section 9(5).
25 Wireless Telegraphy Act 2006, section 8(2).

wireless telegraphy stations or apparatus from requiring a licence[26] and, where OFCOM is satisfied that stations or apparatus of a certain class are unlikely to interfere unduly with other wireless telegraphy networks and services, it must grant an exemption for that class[27]. Examples of apparatus that has been made licence exempt include Bluetooth and WiFi equipment[28]. OFCOM has also granted licence exemptions for mobile phones connected to terrestrial networks, social alarms, hearing aids, 'walkie-talkies' and meter reading and asset tracking devices. Other exemption regulations relate to specific equipment. For example, under the Wireless Telegraphy (Radio Frequency Identification Equipment) (Exemption) (Amendment) Regulations 2007, OFCOM exempted certain radio frequency identification devices from WTA 2006 licensing requirements, in compliance with EC Decision 2006/804, on the basis that they contain no energy source of their own and are therefore unlikely to interfere with other wireless telegraphy apparatus.

In its 2007 Licence-Exemption Framework Review consultation, OFCOM reiterated that:

> 'spectrum use should be licence-exempt if the value that is expected to be derived from the spectrum under such an approach is predicted to be greater than if spectrum use were licensed.'[29]

OFCOM has also stated that it makes sense to consider treating specific frequencies as licence-exempt if the potential for harmful interference is low.

Spectrum may be made available on a licence-exempt basis for use by either specific or multiple applications. However, OFCOM has announced that wherever possible, spectrum it releases for licence-exempt applications in the future will allow for use by multiple applications (what OFCOM refers to as the 'spectrum commons' model). In both cases users are required to comply with certain requirements, principally to operate within specified power limits.

7.3.4 Licence Conditions

OFCOM can grant a wireless telegraphy licence subject to whatever terms, provisions and limitations it sees fit[30]. The limitations may relate to: (i) the proposed usage, users and location of the wireless telegraphy station or apparatus; (ii) the apparatus that may be installed or used; and (iii) in the case of a licence to establish a station only, the position and nature of the station[31]. OFCOM may also impose terms, provisions and limitations relating to: (i) strength or type of signal, times of use and frequency sharing; (ii) content that cannot be broadcast; and (iii) by contrast, requirements for the transmission of particular content by the licensee[32].

---

26 Wireless Telegraphy Act 2006, section 8(3).
27 Wireless Telegraphy Act 2006, section 8(4)–(5).
28 The Wireless Telegraphy (Exemption) Regulations 2003 as amended specifies all licence exempt equipment and the parameters within which that equipment may be used.
29 Introduction to the statement on the framework for managing spectrum used by licence-exempt devices (4 December 2007).
30 Wireless Telegraphy Act 2006, section 9(1). Note that section 9(7) requires that OFCOM may only impose terms, conditions or limitations if it is satisfied that they are objectively justifiable, non-discriminatory, proportionate and transparent.
31 Wireless Telegraphy Act 2006, section 9(2)–(3).
32 Wireless Telegraphy Act 2006, section 9(4).

The conditions attached to a wireless telegraphy licence are listed in the schedules to that licence. These conditions outline the technical parameters imposed on all licensees with the same class of licence and can detail other non-standard conditions (for example, holders of third generation (3G) mobile phone licences were obliged to provide 3G network coverage to 80 per cent of the population by the end of 2007).

Despite the relative freedom afforded to OFCOM to specify licence terms, the terms, provisions and limitations imposed by a WTA licence cannot duplicate general conditions already imposed under section 45 of the Communications Act[33].

7.3.5   Classes and Types of Wireless Telegraphy Licences

The procedure for wireless telegraphy licences is set out in Schedule 1 to the WTA 2006. There are approximately 30 different classes of wireless telegraphy licence, distributed across 12 categories: aeronautical radio; amateur radio; broadcasting and associated services; business radio; fixed terrestrial links; low power and short range devices; maritime radio; mobile and broadband; programme making and special events; radar level gauges; and satellite earth[34]. As might be expected, the licence application and award processes, licence terms and fees payable vary significantly depending on the nature of spectrum use that the licence permits.

Each licence, which consists of a numbered licence document, a set of general terms and conditions applicable to the licence type, and a licence schedule (or schedules), authorises the named licensee (or licensees) to install or use particular radio equipment in a clearly defined way for a defined period of time. The licence schedule lists the technical parameters of allowable use, and includes some standard provisions applicable to the relevant licence class.

Any given licence will be either 'pre-packaged', ie off-the-shelf, generally with no specific assignment or co-ordination required—(an example being a licence for amateur radio) or 'customised', where there is a requirement for individual assignment or a need for co-ordination with other services to reduced the possibility of interference—(for example, a licence for operation of fixed links).

7.3.6   Application and Award of Wireless Telegraphy Licences

Most classes of wireless telegraphy licence are issued directly by OFCOM, with application forms available on OFCOM's website[35]. However, OFCOM has outsourced management of the application process for some licence types to third parties, either for administrative efficiency (where licence grant involves large volumes and little or no discretion—for example, amateur licences), or because decision making requires particular expertise (for example, in the areas of programme making and aeronautical use).

---

33   Wireless Telegraphy Act 2006, section 9(6)–(7).
34   A full list of licence types, with links to information about licensing policy, application processes and licence terms is available at http://www.ofcom.org.uk/radiocomms/ifi/licensing/classes/. An overview of WTA licensing, with relevant links is set out in OFCOM's Wireless Telegraphy Act Licensing Policy Manual, available at http://www.ofcom.org.uk/radiocomms/ifi/licensing_policy_maual_2/. Note that although the list of uses refers to Citizen's Band Radio, the requirement to hold a WTA licence was revoked in December 2006.
35   Application forms can generally be accessed by clicking on the relevant licence class—see link above.

The WTA 2006 requires OFCOM to process licences or further grants of rights to use frequencies within six weeks, except in cases which involve international co-ordination. In addition, OFCOM intends to publish targets for particular licence types, from time to time.

7.3.7 Licence Holders

Licences may be issued in the names of individuals or corporate bodies. Care needs to be taken where there are changes in the names or configuration of persons named on any licence, as this may require the licence to be reissued[36].

Most licences are renewable annually on the payment of a renewal fee.

7.3.8 Publication of Licence Information

OFCOM maintains details of licences granted on the Wireless Telegraphy Register, hosted on its online Spectrum Information System[37]. Broadly speaking, sensitive information, including details of the licensee and relevant technical parameters, will be regarded by OFCOM as confidential. However there are exceptions, in particular where the Freedom of Information Act 2000 applies, or where the licensing process is subject to requirements of transparency[38].

Practical Issue—Sitefinder

OFCOM also operates a database, known as Sitefinder, that provides information on the location and operational characteristics of mobile network base stations. Sitefinder was created following recommendations made by the Independent Expert Group on Mobile Phones in 2000 (commonly referred to as the Stewart Report). Sitefinder is an internet based resource and can be accessed at www.sitefinder.ofcom.org.uk. It is operated by OFCOM on behalf of the Government.

Mobile network operators provide data for inclusion on the Sitefinder basis on a purely voluntary basis. Although the data contained within Sitefinder is owned by the mobile network operators, because OFCOM is a public body all information it holds (including the underlying data provided by the mobile network operators) is subject to the Freedom of Information Act 2000. This was confirmed in a decision by the Information Commissioner and later affirmed by the Information Tribunal. In 2006, a complainant requested that all information OFCOM held regarding based stations be disclosed in various formats. The Information Commissioner held that this included information given to OFCOM by mobile network operators but not actually disclosed on the Sitefinder website. It is fair to say that this decision has had the reverse of the desired effect, as certain mobile network operators no longer provide regular updates to OFCOM. As a result, the Sitefinder database can be out of date.

---

36   See http://www.ofcom.org.uk/radiocomms/ifi/licensing_policy_manual_2/who.
37   The Wireless Telegraphy (Register) Regulations 2004, as amended, require OFCOM to maintain a register of 'relevant' licensing information for licence types listed in the Schedule to the Regulations. These regulations were initially made pursuant to powers under section 170 of the Communications Act 2003. Corresponding powers are now set out in section 31(1) and (2) and section 122(7) of the WTA. Relevant information is information that relates to the grant, renewal, transfer, variation or revocation of wireless telegraphy licences; or the making, renewal, transfer, modification or revocation of grants of recognised spectrum access. The Spectrum Information System can be accessed at http://spectruminfo.ofcom.org.uk/spectrumInfo/.
38   For example, under the terms of the competitive process for the award of spectrum.

7.3.9 Licence Duration

Wireless telegraphy licences are usually granted for an indefinite period, although certain licences are granted for specific periods (second generation mobile phone licences, for example, expire between 2018 and 2021)[39]. Recent wireless telegraphy licences granted at auction for whole blocks of spectrum have an indefinite duration, with a minimum term of 15 years during which OFCOM 's powers to revoke are limited.

A move away from awarding fixed term licences was raised as a possibility in Lord Carter's Digital Britain Interim Report[40]. The rationale given for this is that whilst mobile operators (specifically providers of 3G services) require greater investment certainty in order to commit capital to developing network capacity and coverage, simply awarding a very long term licence actually discourages investment. Instead, Lord Carter has proposed that existing time limited licences could be made indefinite and subject to administered incentive pricing which would be set at a level that generates a return equivalent to the amount that would otherwise be realised if the relevant spectrum was made available to the market at auction. Administered incentive pricing is covered in more detail in the Licence Fees section below.

7.3.10 Licence Revocation and Variation

While licences are generally subject to revocation or variance by OFCOM (discussed in more detail under the 'enforcement and penalties' paragraph below), wireless telegraphy licences can include restrictions on OFCOM's ability to revoke or vary them unless the licensee consents or one of the grounds specified in the licence apply (e.g. there has been a breach of the terms of the licence)[41]. Irrespective of those restrictions, OFCOM can revoke or vary a wireless telegraphy licence at any time in the interests of national security or for the purpose of securing compliance with the United Kingdom's international obligations. It can do this by giving the licence holder written notice of the revocation or variation[42]. Licences can also expressly specify the extent to which regulations made by OFCOM do or do not impact that licence[43].

Revocation of a licence or the variation of its terms can be effected by giving written notice to the licence holder or by publishing a general notice that applies to licences of the class to which the licence in question belongs[44]. OFCOM must notify the relevant licensee prior to revoking or varying its licence. The notice must set out the reasons for the proposed revocation or variation and specify the period (of at least one month) during which the licensee can make representations about the proposed revocation or variation and, where the proposed revocation or variation

---

39 Paragraph 5 of Schedule 1 of the Wireless Telegraphy Act 2006 provides that a wireless telegraphy licence continues in force, unless previously revoked by OFCOM, for such period as may be specified in the licence.
40 The Digital Britain Interim Report was published in January 2009. A copy of the report can be obtained from http://www.culture.gov.uk/what_we_do/broadcasting/5944.aspx.
41 Wireless Telegraphy Act 2006, Schedule 1 paragraph 8(1)–(2).
42 Wireless Telegraphy Act 2006, Schedule 1 paragraph 8(5). The author is not aware that this power has ever been exercised.
43 Wireless Telegraphy Act 2006, Schedule 1 paragraph 8(4).
44 Wireless Telegraphy Act 2006, Schedule 1 paragraph 6.

is the result of a contravention of a licence term, provision or limitation, comply with that term, provision or limitation[45]. Appeals against the revocation of a wireless telegraphy licence can be brought under the terms of the WTA 2006[46].

7.3.11   Licence Fees

Licence fees are set in one of two ways with the amounts payable generally depending on the type of licence. OFCOM can require licensees to pay initial and, if specified in regulations, ongoing fees for licences with the amounts being set by OFCOM through regulations[47]. Section 13(2) of the WTA 2006 adds that OFCOM may, if it thinks fit with regard to its duties under section 3 of the WTA 2006, prescribe sums greater than those required to recover its costs in connection with its radio spectrum functions.

Requiring licensees to pay an annual fee (known as Administered Incentive Pricing or AIP) is the most commonly used pricing mechanism. With AIP, the spectrum manager (eg OFCOM) sets the level of licence fees based on its determination of what constitutes fair market value for the spectrum that is the subject of the licence (or the opportunity cost associated with use of the spectrum). The idea is that charging for spectrum use on this basis promotes efficient spectrum use as a user with unused spectrum will be incentivised to return it instead of paying the relevant fee. In addition, users are likely to explore means of using spectrum more efficiently in order to reduce the fee payable. AIP is applied to the public and private sectors.[48] Although there is widespread support for AIP as pricing mechanism, there are conflicting views as to the basis on which licence fees should be calculated. More information about fees for particular licence types can be found on OFCOM's website (or, where applicable, the website of the third party appointed by OFCOM to administer the licence application process)[49].

OFCOM also has the power to require applicants to bid for spectrum licences through an auction process in accordance with procedures set out by it in regulations[50]. Auctions are seen as a way of facilitating economic efficiency and fairness, through transparency. OFCOM's stated position is that it will use auctions only selectively and for new and regional services.

Fees payable by wireless telegraphy licence holders must be paid when due and OFCOM can adopt recovery procedures where fees are not paid on time[51].

Section 28 of the WTA 2006 provides that the Secretary of State may make payments to OFCOM in respect of Crown use of the radio spectrum.

---

45   Wireless Telegraphy Act 2006, Schedule 1 paragraph 7(1)–(3). Note that OFCOM has the power to extend or reduce the one month period (Schedule1 paragraph 7(4)–(9)).
46   An explanation of how the appeals process works is set out in the Wireless Telegraphy Act Licensing Policy Manual, available from the OFCOM website.
47   Wireless Telegraphy Act 2006, section 12(1)–(2).
48   At the time of writing, there are plans to significantly increase the price paid by the Ministry of Defence for its use of radio spectrum. Also, in 2007, OFCOM announced that from 2014 digital terrestrial radio and television broadcasters will be required to pay AIP.
49   See http://www.ofcom.org.uk/licensing/applications08/changes/Fees.
50   Wireless Telegraphy Act 2006, section 14.
51   Wireless Telegraphy Act 2006, section 15.

7.3.12   Recognised Spectrum Access

Chapter 2 of the WTA 2006 is concerned with grants of recognised spectrum access (RSA). Where a person is proposing to use or continue using a station or apparatus for wireless telegraphy in circumstances specified in regulations made by OFCOM and that use does not require a wireless telegraphy licence but will involve the emission of electromagnetic energy to facilitate the reception of anything in the United Kingdom or in United Kingdom territorial waters (irrespective of whether the emissions are from within the United Kingdom or from a place outside the United Kingdom), OFCOM may make a grant of RSA to that person[52]. The first grants of RSA made by OFCOM were for radio astronomy purposes. OFCOM's rationale for introducing RSA for radio astronomy was to reduce the number of constraints on use in a number of frequency bands and to free up spectrum, in some cases allowing geographical restrictions to be lifted and in others allowing new services to be introduced.

The effect of a grant of RSA is that OFCOM, in carrying out its spectrum management functions, is required to take into account the existence and terms of the grant to the same extent that it would take into account a wireless telegraphy licence which contains equivalent terms, provisions or limitations[53]. It complements the licensing regime in that OFCOM is required to take account of a grant of RSA on a comparable basis to licensed use.

OFCOM can set conditions and restrictions for the grant of RSA as it sees fit, including, restrictions or conditions as to frequency sharing, times of use and strength or type of signal, although these conditions and restrictions cannot duplicate those already imposed under section 45 of the Communications Act[54].

While grants of RSA are generally subject to revocation or modification by OFCOM, grants can, like wireless telegraphy licences, include restrictions on OFCOM's ability to revoke or modify them without, for example, the consent of the holder of the grant or on the grounds specified in the grant itself[55]. Irrespective of those restrictions, OFCOM can revoke or modify a grant of RSA in the interests of national security or public safety or to secure compliance with the United Kingdom's international obligations[56].

Fees for the grant of RSA are set in much the same way as fees for WTA licences with the beneficiaries of RSA grants either paying initial and, if required, ongoing fees or bidding for grants via an auction process[57]. Again, OFCOM may, if it thinks fit in light of the general principles for spectrum allocation, prescribe sums greater than those it requires to recover costs incurred by it in connection with its radio spectrum functions and can bring recovery procedures to recoup fees not paid when due[58].

Section 27 of the WTA 2006 allows OFCOM to make regulations providing for the conversion, on the application of the licence holder, of a wireless telegraphy licence

---

52   Wireless Telegraphy Act 2006, section 18(1)–(3).
53   At the time of writing RSA is only available for radio astronomy.
54   Wireless Telegraphy Act 2006, section 18(6)–(7).
55   Wireless Telegraphy Act 2006, Schedule 2 paragraph 7(2).
56   Wireless Telegraphy Act 2006, Schedule 2 paragraph 7(4).
57   Wireless Telegraphy Act 2006, section 21(1) and section 23.
58   Wireless Telegraphy Act 2006, section 22(2) and section 24.

into a grant of RSA and vice versa. This enables holders of RSA to acquire spectrum licences through the spectrum trading regime and convert them into RSA, thereby extending their RSA holdings. It also allows holders of RSA which only use the facility for certain limited periods of time (eg for one or two months each year) to convert the RSA into spectrum licences and make available other services during the period in which the facility is not used.

7.3.13   Limitations on Authorised Spectrum Use

If OFCOM considers it appropriate to do so, it must make an order, imposing limitations on the use of particular frequencies for the purpose of securing the efficient use of the radio spectrum. An order may:

- specify frequencies for the use of which OFCOM will issue a limited number of wireless telegraphy licences and RSAs only; and/or

- specify uses for which, on specified frequencies, OFCOM will issue a limited number of wireless telegraphy licences and RSAs only.

The order must set out the criteria which OFCOM will apply in determining the limit on the number of wireless telegraphy licences and grants of RSA to be granted and the persons to whom they will be granted. OFCOM must satisfy itself that any such criteria are objectively justifiable, not discriminatory against particular persons, proportionate and transparent in relation to what they are intended to achieve[59].

7.3.14   Regulations Under the WTA 2006

Section 45 of the WTA 2006 grants OFCOM the broad power to make regulations 'prescribing the things that are to be done, or not done,' in connection with the use of wireless telegraphy apparatus or stations. OFCOM can use regulations to impose obligations on licence holders, or those persons in possession or control of wireless telegraphy apparatus, to allow that equipment to be inspected; produce their licence when required; and keep their equipment in a certain condition[60]. Regulations can also be made requiring licence holders to keep and produce accounts and records regarding sums which are or may become due and display notices specified in the regulations at the wireless telegraphy station[61].

In addition to the broad powers to make regulations in section 45, the WTA 2006 lists a range of more specific situations in which regulations can be made. For example, OFCOM can make regulations (which must be approved by the Secretary of State) regarding the use, sale, hire or advertising of certain wireless telegraphy apparatus that generates electromagnetic energy of not more that 3000 GHz to ensure that it does not cause undue interference[62].

OFCOM has used its extensive powers to make regulations in several ways. Among the most common regulations are those that amend licence charges for specific

---

59   Wireless Telegraphy Act 2006, section 29.
60   Wireless Telegraphy Act 2006, section 45(4).
61   Wireless Telegraphy Act 2006, section 45(5)–(6).
62   Wireless Telegraphy Act 2006, section 54.

types of licence holders[63] and those that amend the conditions attached to existing licences, for example by authorising licensees to transfer certain rights and obligations under their licences[64].

7.3.15 Spectrum Trading

Section 3 of the WTA 2006 allows OFCOM to make regulations authorising wireless telegraphy licence holders and the holders of a grant of RSA to transfer rights and obligations under those licences or grants to a third party. Spectrum trading was first introduced in the United Kingdom by the Wireless Telegraphy (Spectrum Trading) Regulations 2004, as amended (the Spectrum Trading Regulations).

The Spectrum Trading Regulations allow for three different types of transfer:

- outright transfers—where all the rights and obligations under a licence are transferred from one party to another (so that the transferor no longer has any rights or obligations under the licence once the transfer is completed);

- concurrent transfers—where the rights and obligations under a licence are transferred on a concurrent basis (so that the transferor and transferee both share the rights and obligations under the licence once the transfer is completed); and

- partial transfers—where only some of the rights and obligations under a licence are transferred from one party to another (so that the rights and obligations are held under two distinct licences once the transfer is completed). Partial transfers can be outright or concurrent and the transfer can be on a frequency, geographical or time basis.

Trading may only take place in licence classes that have been designated as tradeable and the type of trade that can be carried out varies depending on the whether the relevant licence class allows for all or only some types of trade[65].

Companies that hold a tradeable wireless telegraphy licence can transfer the rights and obligations under that licence to a third party. The transfer process is set out in regulations 8, 9 and 10 of the Spectrum Trading Regulations[66].

---

63 For example, the Wireless Telegraphy (Licence Charges) (Amendment) Regulations 2007 altered the fees for a number of wireless telegraphy licences and two further amendments affecting charges have been made in 2008.

64 For example, the Wireless Telegraphy (Spectrum Trading) (Amendment) (No 2) Regulations 2008 authorised spectrum trading among five new classes of business radio licences.

65 At the time of writing, a number of classes of wireless telegraphy licences are transferable including certain fixed-wireless access, business radio and fixed links licence classes. In early 2008, OFCOM announced that public bodies in the United Kingdom would be able to sell and trade radio spectrum. This means that the Ministry of Defence (which holds around a quarter of the useful spectrum below 15 GHz) will be entitled to sell off or trade highly prized spectrum which suitable for a wide range of uses.

66 OFCOM maintains a Trade Notification Register which details all proposed trades notified to OFCOM, trades in progress and completed trades. This can be accessed at http://spectruminfo.ofcom.org.uk/ spectruminfo/trades. For information on the tax implications of spectrum trading, refer to http:// www.ofcom.org.uk/radiocomms/ifi/trading/tradingguide/tradingguide.pdf.

7.3.16 Enforcement and Penalties

The WTA 2006 contains a range of offences and penalties relating to many aspects of spectrum management. This section is focused on offences that result from the breach of the requirements for, and conditions of, wireless telegraphy licences.

The WTA 2006 states that it is a criminal offence for a person to:

■ establish or use a wireless telegraphy station, or install or use wireless telegraphy apparatus, for the purpose of making a broadcast, without a licence (where a licence is required under section 8 of the WTA 2006)[67];

■ possess or control a wireless telegraphy station or apparatus with the intention of using it, or knowing or having reasonable grounds to believe that someone else intends to use it, without a licence (where a licence is required under section 8 of the WTA 2006)[68];

■ knowingly cause or permit premises, of which that person is in charge, to be used for unlawful broadcasting or to have reasonable grounds to believe that the premises are being so used without taking reasonable steps to prevent them from being so used[69].

The penalty for committing any of these offences is: (i) on summary conviction, up to 12 months imprisonment (six months in Northern Ireland and Scotland), a fine up to the statutory maximum or both; or (ii) on conviction on indictment, up to two years imprisonment, a fine or both.[70]

The WTA 2006 also makes it a criminal offence to install or use receiving apparatus (ie equipment only able to receive broadcasts, not emit them) for receiving unlicensed broadcasts in contravention of sections 35 or 36, an offence which is punishable with a fine not exceeding level three (currently £1,000)[71]. A person who commits any other offence under section 35 or section 36 is liable on summary conviction to up to 51 weeks imprisonment (six months in Northern Ireland and Scotland), a level five fine (currently £5,000) or both[72].

While the offences above are targeted at unlicensed broadcasters and, in the case of receiving equipment, consumers of unlicensed broadcasts, the WTA 2006 lists numerous activities, broadly summarised as facilitating unauthorised broadcasts, that also constitute offences. For example, it is an offence knowingly to participate in the day-to-day running or financing of a broadcast station making unauthorised broadcasts; to supply, install or maintain equipment used in making unauthorised broadcasts; or to participate in an unauthorised broadcast, for example as a performer[73].

The WTA 2006 also specifies the procedure that OFCOM needs to follow when licence holders, or people operating under section 8(3) exemptions, fail to comply with the terms of their licences or licence exemptions. Where OFCOM determines

67 Wireless Telegraphy Act 2006, section 35(1)–(2).
68 Wireless Telegraphy Act 2006, section 36(1).
69 Wireless Telegraphy Act 2006, section 37(1).
70 Wireless Telegraphy Act 2006, section 35(2)–(3), section 36(2)–(3) and section 37(2)–(3).
71 Wireless Telegraphy Act 2006, section 35(4) and section 36(4).
72 Wireless Telegraphy Act 2006, section 35(5) and section 36(5).
73 Wireless Telegraphy Act 2006, section 38(2).

that there are reasonable grounds for believing that a person is contravening or has contravened a term, provision or limitation of a wireless telegraphy licence or licence exemption, OFCOM must notify the person specifying, among other things, the deadline for that person to make representations about the matter and/ or comply with the term, provision or limitation being contravened[74]. This deadline must be one month from the day after the one on which the notification was given although OFCOM can: (i) if it thinks fit, specify a longer deadline or extend the initial (or any subsequent) deadline; or (ii) specify a shorter deadline if OFCOM has reasonable grounds for believing that the case is a case of repeated contravention and the licensee is notified of the shorter deadline[75]. A repeated contravention is deemed to have taken place where the person in question has received a previous notification under section 39 or been convicted of an offence under section 35 in the 12 months before the subsequent notification[76].

OFCOM may not bring proceedings under section 35 for contravening a wireless telegraphy licence or licence exemption: (i) before it has notified the person in question (the defendant) under section 39 and considered any representations made by the defendant within the relevant deadline; and (ii) if the defendant subsequently complies with the notification by the relevant deadline[77]. However, it should be noted that OFCOM can bypass section 39 if it certifies that the procedure is inappropriate because of an immediate and serious risk to public safety, health or national security or of economic or operational problems for other people using wireless telegraphy stations or apparatus or communications providers[78].

Fines can also be imposed for breaching regulations made under the WTA 2006[79] although the threat of fines and/or imprisonment is not the only enforcement weapon available to OFCOM. OFCOM can revoke, or vary the terms, provisions or limitations of a licence and revoke or modify a grant of RSA, or the conditions or restrictions to which it is subject, by notifying the licensee and allowing him a period to make representations about the proposal or comply with the licence or grant of RSA, where OFCOM's notice has been prompted by a contravention.

The threat of varying a licence has been used recently by OFCOM against O2. When O2 failed to meet the obligation in its licence to provide 3G network coverage to 80 per cent of the population by the end of 2007, OFCOM issued a notice under the WTA 2006 threatening to reduce the term of O2's 3G licence by four months (estimated to be equivalent to a financial sanction of at least £40 million) if the coverage obligation was not met by 30 June 2008.

It should be noted that the WTA also outlines a range of offences for the misuse of wireless telegraphy. Sending false or misleading messages likely to disrupt a 'safety of life service' or endanger the safety of a person, ship, aircraft or vehicle, using wireless telegraphy apparatus deliberately to intercept messages intended for another recipient, or using apparatus for the purpose of interfering with wireless telegraphy, constitute criminal offences[80].

74 Wireless Telegraphy Act 2006, section 39(1)–(3).
75 Wireless Telegraphy Act 2006, section 39(5)–(7).
76 Wireless Telegraphy Act 2006, section 40.
77 Wireless Telegraphy Act 2006, section 41(1)–(3).
78 Wireless Telegraphy Act 2006, section 41(4).
79 Wireless Telegraphy Act 2006, section 46.
80 Wireless Telegraphy Act 2006, sections 47, 48 & 68.

Finally, it is worth noting four general points in relation to the enforcement of WTA 2006 offences.

- First, in enforcing the WTA 2006 (other than Part 4 or section 111), a justice of the peace can grant a search warrant where there is a reasonable ground for suspecting that (subject to certain limited exceptions) an offence under the WTA 2006 has been, or is being committed, and that evidence for the offence will be found on the premises, vehicle, ship or aircraft specified in the warrant[81]. In most cases where a search warrant has been granted, it can authorise the seizure of wireless telegraphy apparatus or any other item that appears to be relevant to the investigation[82].

- Secondly, those in charge of organisations whose activities fall within the auspices of the WTA 2006 should be aware that where an offence is committed by a body corporate but it takes place with the consent or connivance of, or is attributable to the neglect of, a director, manager, secretary or similar officer, or person purporting to act in such capacity, that person is liable to be prosecuted as well as the body corporate[83].

- Thirdly, where a person is convicted of a relevant offence under sections 35–53 inclusive[84], 66 and 68 (and certain other sections of the WTA 2006), the court may, in addition to any other penalty, order wireless telegraphy apparatus, other related equipment, vehicles, ships, aircraft or other objects connected to the offence to be forfeited to OFCOM[85].

- Finally, OFCOM can bring civil proceedings in relation to a contravention which is an offence under the WTA 2006 for an injunction or any other appropriate relief[86].

7.4 OFCOM's Management of the Radio Spectrum

The current regulatory approach to managing spectrum in the United Kingdom is based on proposals laid out by OFCOM in its Spectrum Framework Review, published in November 2004. In the Spectrum Framework Review, OFCOM set out its intention to gradually move away from the prevailing 'command and control' approach to spectrum management towards a more market based system which allows the market to decide what particular frequencies should be used for (within parameters set by OFCOM). The rationale for this policy shift was that the market is best positioned to ascertain the value of radio spectrum and freeing spectrum of technology and usage constraints that are mandated by the regulator should aid the efficient development and roll-out of innovative technologies and services.

7.4.1 OFCOM's Spectrum Management Vision

In the Spectrum Framework Review, OFCOM summarised its vision for spectrum management as follows:

---

81 Wireless Telegraphy Act 2006, section 97(1).
82 Wireless Telegraphy Act 2006, section 99(2).
83 Wireless Telegraphy Act 2006, section 110.
84 Other than offences relating to receiving apparatus under sections 35 and 36 or under section 51(4).
85 Wireless Telegraphy Act 2006, Schedule 5, paragraph 1.
86 Wireless Telegraphy Act 2006, section 108(2).

- 'Spectrum should be free of technology and usage constraints as far as possible. Policy constraints should only be used where they can be justified;

- It should be simple and transparent for licence holders to change the ownership and use of spectrum; and

- Rights of spectrum users should be clearly defined and users should feel comfortable that they will not be charged without good cause.'[87]

OFCOM 's preference for a market based approach to spectrum management stems, in part, from the work carried out by Professor Martin Cave for the Department of Trade and Industry and the Treasury in early 2002 and the resulting proposals documented in his report (the Cave Report)[88]. OFCOM 's shift towards a looser spectrum management policy should be understood in the context of the recommendations made in the Cave Report, principally that a market-led approach to spectrum management would result in the increasingly efficient and innovative use of radio spectrum in the United Kingdom.

To achieve its vision for spectrum management, in the Spectrum Framework Review OFCOM proposed:

- providing spectrum for licence-exempt use as required (although OFCOM does not anticipated this will exceed more than seven per cent of the total radio spectrum);

- introducing a market based approach to spectrum management through the implementation of spectrum trading and liberalisation where possible. OFCOM believes that it can fully implement these policies in around 72 per cent of the radio spectrum; and

- managing the remaining 21 per cent of the radio spectrum using a 'command & control' approach where OFCOM allocates and assigns the spectrum.

In the Spectrum Framework Review: Implementation Plan[89], OFCOM identified that the transition away from a command and control approach to spectrum management to an increasingly deregulated regime should be achieved by the development and implementation of three specific policies: spectrum trading (covered in section 7.3 above), spectrum liberalisation and releasing spectrum into the market as soon as possible after it becomes available, with minimum restrictions on how it can be used.

7.4.2 Spectrum Liberalisation

Spectrum liberalisation is the term used to describe the reduction or removal of technology restrictions from wireless telegraphy licences (or at least those that aren't strictly necessary to prevent undue interference to other users or as required by international agreements), so that the frequencies to which a licence relates can be used for the provision of other services and with different technologies. It was first introduced in the United Kingdom in January 2005 in the Business Radio, Fixed Wireless Access and Fixed Links sectors.

---

87 Spectrum Framework Review, OFCOM consultation document, 23 November 2004, section 1.7.
88 http://www.ofcom.org.uk/static/archive/ra/spectrum-review/index.htm.
89 Spectrum Framework Review: Implementation Plan, OFCOM consultation document, 13 January 2005.

Along with spectrum trading, spectrum liberalisation is one of the key mechanisms identified by OFCOM to facilitate the move away from the traditional command and control approach to spectrum management to a more market orientated approach. In OFCOM's view, spectrum liberalisation helps ease restrictions which serve as a barrier to market entry for new service providers. As a consequence, it should become easier for users to deploy new and innovative technologies and applications, thereby increasing competition. However, proposals for the introduction of spectrum liberalisation in the United Kingdom have not been without controversy, particularly in relation to proposals to remove restrictions on the use of spectrum for mobile services and the liberalisation of spectrum presently used for 2G services. Indeed, in his Digital Britain—Interim Report, Lord Carter, recognising that the current impasse between the mobile network operators and OFCOM over the future use of existing 2G radio spectrum is preventing the release of critical spectrum into the market, proposed that existing 2G licensees should be allowed to re-align their existing spectrum holdings and re-use the 2G spectrum allocated to them for the deployment of next generation mobile services (including 3G services). Lord Carter set a deadline of the end of April 2009 for the mobile industry to enter into a voluntary industry-wide consensus with OFCOM. Although claiming that failure to do so would most likely result in the Government supporting 'an imposed solution', at the time of writing this deadline had passed and no action had been taken by the Government.

7.4.3 The Digital Dividend Review

Described by OFCOM as 'one of the most important decisions we have ever made', the digital dividend review (DDR) is the process by which OFCOM is deciding how to award the radio spectrum freed up by the switchover from analogue television broadcasting to digital terrestrial television (DTT). Consultation on initial proposals was launched in December 2006, with a formal statement of OFCOM's approach following in December 2007[90].

The spectrum to be released is in the UHF band in the frequencies 470–862 MHz. There are three types of spectrum covered by the DDR: cleared spectrum; spectrum with potential for clearance (currently used for airport radar and wireless microphones); and interleaved spectrum (i.e. the 'white space' that exists geographically between television transmitters to prevent interference).

OFCOM is, in contrast with its historic approach to spectrum regulation, taking a 'market led approach' to awarding these classes of spectrum, meaning that, with one important exception, it will auction the spectrum in a manner that will allow spectrum holders to decide how the relevant spectrum will be used, rather than mandating the purposes of its use. OFCOM sees reservation of spectrum for particular uses as reducing flexibility and the opportunity to introduce new technologies and services. It believes that the market is better placed than the regulator to determine the best uses of spectrum.

However, there are some important exceptions to OFCOM's general approach as demonstrated by its proposed treatment of PMSE (the 'programme management and special events' sector), which already uses interleaved spectrum on a large scale.

---

90 'Digital Dividend Review – A statement on our approach to awarding the digital spectrum', OFCOM statement, 13 December 2007.

OFCOM does not believe the PMSE sector would be able to take part effectively in an auction process, and plans to hold a beauty contest to award a package of interleaved spectrum to meet PMSE users' needs. OFCOM intends to appoint a band manager for a package of interleaved spectrum, who will be required to give access to PMSE users on regulated terms, but otherwise entitled to allow others to use the relevant spectrum.

Originally, OFCOM anticipated holding an auction for the cleared spectrum in the summer of 2009. However, as a result of important new events since June 2008—principally the decisions taken by a large number of European countries to include the 800 MHz spectrum band in their digital dividend and moves by Europe to harmonise use of this spectrum—in February 2009, OFCOM published a consultation document in which it indicated that it expects to hold an auction for the cleared spectrum in 2010[91].

---

91   http://www.ofcom.org.uk/consult/condocs/800mhz/800mhz.pdf.

Chapter 8

Powers and Duties

Alex Haffner　　　　　　　DentonWildeSapte...
Denton Wilde Sapte LLP

1 Introduction

1.1 OFCOM

The Office of Communications Act 2002 formally established OFCOM as the regulator for the UK communications industries, although it was not until the passing of the Communications Act 2003 (Communications Act) that it took on the powers and duties it now exercises and which form the basis for this chapter[1]. Broadly speaking, Part 1 of the Communications Act sets out OFCOM's functions and specifies its duties, whereas Part 2 contains the detailed provisions on its powers in respect of network services and the radio spectrum[2].

The Communications Act reflects an acknowledgement by central government that convergence between content and distribution networks necessitated the unification of the separate 'legacy' regulators for telecoms, television and radio[3]. It was thereby intended that OFCOM should, in the future, make decisions on a converged and integrated basis.

1.2 OFCOM's Relationship with the Secretary of State

OFCOM is, legally, a 'statutory corporation'. It does not report directly to Parliament but, in practice, it is subject to oversight in the performance of its powers and duties by central government through the Department of Culture Media and Sport (DCMS) and the Department of Business Enterprise and Regulatory Reform (DBERR). Members of the OFCOM Board (see Section 3 below) are formally appointed by the Secretary of State from each of these Departments (referred to in the remainder of

---

1　Provisions relating to the management of the radio spectrum and the new regulatory regime for electronic communications entered into force on 25 July 2003 (to meet the EU deadline for implementation of the Framework Directives), but OFCOM did not take on all its powers and duties until 29 December 2003.

2　Also of relevance in the context of this chapter are OFCOM's powers in respect of preserving competition in communications markets contained in Part 5 of the Act.

3　Telecoms regulation was previously undertaken by the Office of Telecommunications (Oftel) with specific powers of enforcement granted to the Director-General of Telecommunications under the Telecommunications Act 1984.

this Section in the singular), who also act in partnership with OFCOM on a number of matters of perceived national or international importance[4]. For example, one of the most important changes implemented under the Communications Act 2003, was the replacement of the previous system of individual licences given to communications providers (issued under the Telecommunications Act 1984) with a general authorisation regime. As part of this new regime (discussed in more detail at chapter 3) OFCOM has the power to require the providers of certain designated types of electronic communication networks (ECNs) or electronic communications services (ECSs) to notify it before they begin operations. Designation for this purpose is undertaken by OFCOM, but expressly subject to consultation with the Secretary of State[5] although, to date, no such designations have in fact been made.

With the passing of the Communications Act, OFCOM took on for itself a number of powers and duties previously exercised directly by the Secretary of State. Nevertheless, the Act still leaves the Secretary of State with certain residual powers, particularly in those areas of regulatory enforcement where it is considered that the 'public interest' requires intervention by central government. These include the power:

- to give directions in respect of network or spectrum functions (Communications Act 2003, section 5)—see Section 2.2 below;

- by order, to set out those networks and services which should be subject to universal service conditions (Communications Act, section 65), the designation of providers of those services and their compliance with the universal service conditions being undertaken by OFCOM—see section 5 below[6]; and

- to require the suspension or restriction of any ECN or ECS providers' entitlement to continue operating their network or system (Communications Act, section 132)—see Section 8(A)(iv) below.

The Secretary of State is under a duty to prepare and lay before Parliament an 'annual report' as to the exercise of their functions under the Communications Act, Wireless Telegraphy Act 2006, Office of Communications Act 2002 (see above) and The Broadcasting Acts 1990 and 1996[7].

2 Duties Under the Communications Act

OFCOM's regulatory powers are considerable, being those previously carried out by the Secretary of State and the legacy regulators, in addition to the powers specifically conferred on it by the Communications Act. With those powers come a wide range of duties. OFCOM itself has calculated it has 263 statutory duties[8], compared to 128 imposed on the five legacy regulators it placed. These duties are often relied on by aggrieved parties to challenge OFCOM decisions. In particular,

---

4   The Secretary of State also has a specific role in ensuring that OFCOM complies with all applicable European law.
5   Communications Act 2003, sections 33 and 34 (see, especially, Section 34(2)).
6   In theory, therefore, it would be up to the Secretary of State to specify any broadband universal service obligation, a topic which is the subject of much current debate.
7   Communications Act 2003, section 390. See, for example, the *Fourth Report on the Secretary of State's functions* (HC 1037), October 2008.
8   See, eg *A case study on public sector mergers and regulatory structures* (April 2006), at page 4. Available at www.ofcom.org.uk/about/accoun/case-study/publicsector.

the Competition Appeal Tribunal (CAT), which has jurisdiction to hear a wide range of appeals, has become a forum for the consistency of OFCOM's approach to be tested against the duties imposed on it by statute (see further at Section 9 below).

2.1 General Duties

The Communications Act specifies two principal duties which define OFCOM's role in carrying out its functions:

- to further the interests of citizens in relation to communications matters; and

- to further the interests of consumers in markets within its areas of responsibility, where appropriate by promoting competition (section 3(1)).

It can be seen therefore that OFCOM's remit is to pay particular regard to consumer choice backed up by effective competition.[9] In the telecommunications sector it must carry out those principal duties with a view to securing:

- optimal use of the electromagnetic spectrum; and

- the availability throughout the UK of a wide range of electronic communications services (section 3(2)).

Also in relation to telecommunications regulation, section 3(4) requires OFCOM to have regard, when performing its duties, to each of the following as appear relevant in the circumstances:

- the desirability of promoting competition in relevant markets;

- the desirability of promoting and facilitating the development and use of effective forms of self-regulation[10];

- the desirability of encouraging investment and innovation in relevant markets;

- the desirability of encouraging the availability and use of high-speed data transfer services throughout the United Kingdom;

- the different needs and interests, so far as the use of the electromagnetic spectrum for wireless telegraphy is concerned, of all persons who may wish to make use of it;

- the vulnerability of children and others whose circumstances appear to OFCOM to put them in need of special protection; and

- the desirability of preventing crime and disorder.

Where a conflict between these duties and OFCOM's Community requirements (see Section 2.2 below) arises, priority must be given to the Community requirements (section 3(6)). Otherwise, where any particular general duty conflicts with another such duty, OFCOM must resolve the conflict in the manner it thinks best in the circumstances (section 3(7)).

---

9 A report by Consumer Focus, published on 26 February 2009, into the performance of six regulators against a series of consumer-focused indicators, specifically praised OFCOM for its *genuine commitment to meeting the needs of citizens and consumers throughout the whole of the UK*. See *Rating Regulators*, available at www.consumerfocus.org.uk.

10 On 10 December 2008, OFCOM published a statement on the general principles it will refer to when determining the likely effectiveness of self and co-regulation in meeting its objectives.

2.2 Directions by the Secretary of State in Respect of Network or Spectrum Functions

As noted above, although OFCOM is able to exercise many of its powers independently of Government, this is expressly subject to the ability of the Secretary of State to intervene in the public interest. To that end, the Secretary of State may issue directions to OFCOM as to the exercise of its powers under Part 2 of the Communications Act or those enactments relating to the management of spectrum (see Chapter 7 above) for one of the purposes set out in Section 5(3), namely:

- in the interests of national security;

- in the interests of relations with other countries;

- to secure compliance with international obligations; or

- for public safety or health reasons.

A direction issued in this manner cannot suspend or restrict a person's entitlement to provide an electronic communications network or service or any associated facilities (section 5(4))[11].

To date, these powers do not appear to have been exercised by the Secretary of State.

2.3 Community Duties

As explained in the earlier chapters of this book, UK telecoms regulation takes place within the context of the European Framework for Electronic Communications. The Communications Act fulfils the UK's obligations in respect of four of the directives contained in the existing framework[12]. OFCOM is also required to act in accordance with six 'Community requirements' which give effect, amongst other things, to the requirements of article 8 of the Framework Directive. The Community requirements are:

- to promote competition (section 4(3));

- to ensure that its activities contribute to the development of the European internal market (section 4(4));

- to promote the interests of all persons who are citizens of the European Union (section 4(5));

- take account of the desirability of carrying out its functions in a manner which, so far as practicable, does not favour one form of network, service or associated facility, or one means of providing or making available such a network, service or facility over another (section 4(6));

- to encourage the provision of network access and service interoperability (section 4(7))[13]; and

---

11   Note, however, that this does not affect the Secretary of State's powers to require suspension or restriction of a provider's entitlement under Section 132 (see Section 8.1.4 below).

12   Directive 2002/21/EC (the Framework Directive), Directive 2002/20/EC (the Authorisation Directive), Directive 2002/19/EC (the Access and Interconnection Directive), and Directive 2002/22/EC (the Universal Services Directive).

13   'Network access' and 'service interoperability' are each defined in Section 151 of the Communications Act.

- to encourage compliance with international standards to the extent necessary to facilitate service interoperability, and to secure a freedom of choice for customers (section 4(9)).

2.4 Other Regulatory Principles

It is OFCOM's publicly stated objective to operate with a bias against regulatory intervention. Correspondingly, Part 1 of the Communications Act contains further duties as to the transparency of OFCOM's enforcement powers and the need to ensure, where possible, that it adopts a 'light touch' approach to regulation. Section 3(3) of the Communications Act states that in all cases OFCOM must have regard to:

'*(a) the principles under which regulatory activities should be transparent, accountable, proportionate, consistent and targeted only at cases in which action is needed; and*

(b) any other principles appearing to OFCOM to represent the best regulatory practice'.

OFCOM has set out its interpretation of these principles as follows:

Practical Issue: OFCOM's Regulatory Principles[14]

- OFCOM will regulate with a clearly articulated and publicly reviewed annual plan, with stated policy objectives.
- OFCOM will intervene where there is a specific statutory duty to work towards a public policy goal which markets alone cannot achieve.
- OFCOM will operate with a bias against intervention, but with a willingness to intervene firmly, promptly and effectively where required.
- OFCOM will strive to ensure its interventions will be evidence-based, proportionate, consistent, accountable and transparent in both deliberation and outcome.

- OFCOM will always seek the least intrusive regulatory mechanisms to achieve its policy objectives.
- OFCOM will research markets constantly and will aim to remain at the forefront of technological understanding.
- OFCOM will consult widely with all relevant stakeholders and assess the impact of regulatory action before imposing regulation upon a market.

To ensure that it meets these objectives, the Communications Act imposes further obligations on OFCOM as to:

Review of the regulatory burden (Section 6): OFCOM must review its functions to ensure that any regulation does not lead to the imposition or maintenance of burdens which are or have become unnecessary. The Communications Act provides

---

14 These were first published in OFCOM's *Foundation and Framework document* (September 2003).

that it should publish a statement setting out how it proposes to comply with this duty (Section 6(4))[15].

Impact assessments (Section 7): Other than where it is impractical or inappropriate for it to do so in view of the urgency of the matter, before implementing an 'important proposal', OFCOM must either carry out and publish an assessment of the likely impact of the proposal or publish a statement setting out why it does not believe it necessary to carry out that assessment. The Communications Act defines an important proposal as one which would involve a major change in OFCOM's activities or have a significant impact on (communications) businesses or the public.

OFCOM's annual report sets out those impact assessments which have been made under this Section, with a summary of the decisions taken on proposals which were the subject of such assessments. OFCOM has also published guidelines on its approach to impact assessments[16].

In the recent case of *Vodafone v OFCOM*, the CAT upheld an appeal by Vodafone against a decision by OFCOM to modify arrangements for number portability (the process by which customers keep their telephone number when they switch service provider). The CAT found that OFCOM had failed to carry out its cost benefit analysis of the changes to the necessary standard. In particular, when it had consulted industry participants, OFCOM had not provided a technical specification on which consultees could provide any useful data[17].

Promptness standards (Section 8): OFCOM is required to publish a statement which sets out the standards it proposes to meet on promptness in carrying out its functions. The time limits set will not apply where the Communications Act (or any other relevant enactment) already sets down a time limit—for example, the time limits which apply on the resolution of disputes under Section 185 of the Communications Act—see Section 6 below.

3 Corporate Structure

By any standards, OFCOM is a large organisation; its budget for the year to 31 March 2010 is £136.8 million[18]. As at 31 March 2008 its headcount was 812 (although this represents a significant drop on the combined headcount of the five legacy regulators which it replaced). A diagram showing OFCOM's basic governance structure is shown in Figure 8.1.

---

15 OFCOM complies with this duty through the publication of a 'Simplification Plan', outlining what steps it has taken to reduce regulation and the further initiatives it intends to undertake in this regard. The latest version of this document was published in December 2008 and is available on OFCOM's website.
16 Better Policy Making: OFCOM's approach to Impact Assessment, issued on 21 July 2005.
17 *Vodafone v OFCOM* [2008] CAT 22.
18 Most of this money is self-generating in that OFCOM imposes 'regulatory fees' on those whom it regulates; any under spend of the annual budget is thereby remitted back to industry in the form of reduced charges (see Communications Act, Section 38).

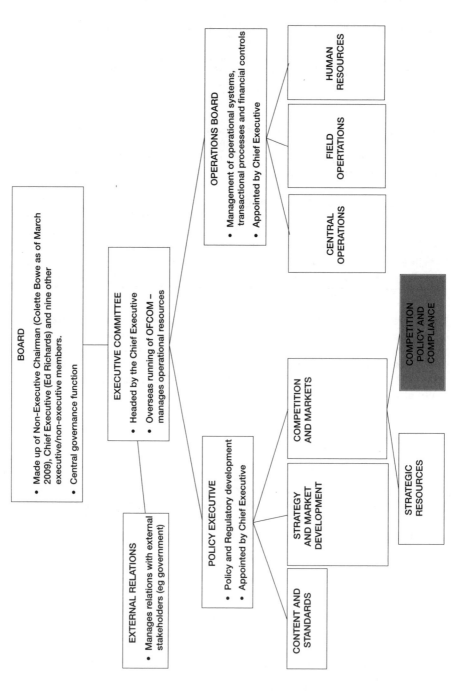

Figure 8.1—OFCOM—Organisational Structure

It can be seen that OFCOM is structured in a similar manner to many commercial enterprises; this was a deliberate attempt to ensure that it was reflective of the entities whom it regulates.

At the apex of the organisation is the Board, which takes overall responsibility for performing OFCOM's powers and duties under the Communications Act and has oversight over funding and expenditure. The Board comprises a non-executive Chairman, Chief Executive and nine other executive and non-executive members who meet formally every month[19]. Several committees are appointed by the Board to discharge certain of its functions and/or advise on issues of relevance to its decision-making powers.

Whereas the Board concentrates on providing strategic direction for OFCOM, day-to-day management of the organisation is the responsibility of the Executive Committee (ExCo) under the leadership of the Chief Executive[20], who is effectively the 'public face' of the organisation. Within ExCo are two further executives: the Policy Executive and the Operations Board. The Operations Board is mainly concerned with managing operational systems and transactional processes within OFCOM, whilst the Policy Executive is divided into three further groups:

- Content and Standards (which deals with broadcast matters including TV/Radio standards, consumer protection and media literacy);

- Strategy and Market Development (dealing with strategy, market intelligence and consumer research); and

- Competition and Markets (dealing with compliance and ownership investigations, licensing and spectrum trading).

Many of OFCOM's competition and regulatory functions are therefore concentrated within the Competition and Markets group. In particular, it is this group which runs OFCOM's Investigations Programme—see further at Section 4 below.

4 Powers of Enforcement: OFCOM's approach to investigations

Before deciding whether to use any of its powers of enforcement, OFCOM will first evaluate the evidence before it carefully to see whether it should dedicate further internal resource to the exercise of those powers. OFCOM's approach to the conduct of investigations, including the principles it applies in 'prioritising' work (see below), are set out in its draft Enforcement Guidelines (Enforcement Guidelines)[21].

Otherwise than in respect of spectrum (the enforcement regime as it applies to spectrum is set out at Chapter 7) and content which have dedicated teams, investigations are undertaken by a specialised unit ('Investigations') within the Competition Group. Broadly speaking, the role of the Investigations Team is to

---

19 Agenda and notes of Board meetings (as well as a number of other executive/committee meetings) are regularly published on the OFCOM website at www.ofcom.org.uk.

20 Ed Richards currently sits in this post.

21 *OFCOM Draft Enforcement Guidelines*: published for consultation on 6 July 2006. Note that, whilst these guidelines are in draft form and a final version is still to be published, OFCOM has stated that it effectively relies on them in the performance of its powers and duties.

identify (and where appropriate act on) areas of concern regarding communications providers' compliance with rules on consumers' rights and competition between communications companies.

OFCOM will undertake investigations on the basis of information received from third parties, or from information it generates itself through the exercise of its powers and duties (so called 'own—initiative investigations').

A company or individual wishing to persuade OFCOM to exercise its enforcement powers will need first to bring the matter to OFCOM's attention in the form of a complaint or dispute submission. Guidance as to what information should be included is set out in the Enforcement Guidelines.[22] When OFCOM receives a complaint/dispute it will first be screened by the 'Operations Team' which sits within the Competition Group to assess whether the matter should be accepted or rejected on receipt (eg because the information provided is incomplete).

Once a matter is accepted by the Operations Team it is classed as an enquiry and passed on to the Investigations Unit which appoints a case team[23] to decide whether to conduct a fuller investigation. This decision is usually taken within 15 working days, although in relation to competition law complaints OFCOM has extended this period to eight weeks. OFCOM has limited resources and therefore inevitably has to prioritise its enforcement activities. This is particularly relevant in the light of those rights of appeal which arise when OFCOM takes any 'decision' other than to close a file on 'administrative' grounds—see further at Section 9.2 below[24].

Where OFCOM opts to conduct a full investigation, it will inform the parties concerned and publish an entry to that effect in its *Competition and Consumer Enforcement Bulletin*[25], which alerts any interested third parties who may wish to participate in the investigation. Thereafter, the focus of the case team's efforts will be on gathering as much information as they judge necessary to reach a final decision. Otherwise than in resolving disputes (see Section 6.2 below), OFCOM is not under any statutory duty to complete its investigation within a fixed time period. Nevertheless, it has target timescales which are set out in its Enforcement Guidelines (and which do not usually extend beyond six months)[26]. To ensure the investigatory process is as efficient as possible, Sections 135–139 of the Communications Act give OFCOM statutory power to seek the information it needs from parties to carry out an investigation and impose penalties for non-compliance with any such request.

5 Enforcement of Conditions

Section 45 of the Communications Act allows OFCOM to set conditions which are either: (a) general conditions; or (b) specific conditions which fall within one of the following four headings:

22 See Annex 4 of the Enforcement Guidelines.
23 The case team will be a variety of personnel from the Investigations team covering (as relevant) appropriate legal, economic, financial and technical expertise.
24 An indication as to what action OFCOM has taken in relation to the complaints and disputes it receives over a given period is provided in its 'Investigations Programme Report' covering activity over the previous six month period.
25 www.ofcom.org.uk/enforcement.
26 Enforcement Guidelines at paragraph 5.43 (Table 1).

- **Universal service conditions**: to ensure compliance with a 'universal service order' made by the Secretary of State (section 67);

- **Access-related conditions**: concerning the provision of network access and service interoperability (sections 73 to 76);

- **Privileged supplier conditions**: relating to public communications providers that enjoy special or exclusive rights in connection with the provision of any non-communications services (section 77); and

- **Significant market power (SMP) conditions**: for markets which have been reviewed by OFCOM and SMP findings made (sections 87 to 93).

General conditions are considered in more detail in Chapter 3 and specific conditions Chapter 4 above.

5.1   Threshold for Intervention and Enforcement Powers

Sections 94–104 of the Communications Act set out OFCOM's powers to ensure compliance with general and specific conditions. The threshold for intervention is where OFCOM has '*reasonable grounds for believing that a person is contravening, or has contravened, a condition*' (section 94(1)). Where, on the basis of its enquiries, OFCOM considers the threshold under section 94(1) has been met, it will issue a notification (Notification) to the person concerned (the notified provider). That Notification sets out the breach and steps required to ensure compliance with the condition(s) at issue within a set period—usually one month. Pursuant to section 98(3), OFCOM may shorten this period in view of the 'urgency' of a particular case[27]. The notified provider will also be given an opportunity to make representations about the matters notified.

Once the period for making representations has expired, OFCOM reviews the steps taken by the notified provider to remedy the consequences of their breach. Assuming that, at this stage:

- the provider is still in contravention of a condition; and

- has not (since the section 94 notice was given) taken all such steps as OFCOM considers appropriate for complying with the condition and correcting the effects of the notified contravention,

then OFCOM will serve a written notice (Enforcement Notice) on the provider setting out those steps which must be undertaken to ensure compliance.

Under section 96, OFCOM is also entitled, when it issues an Enforcement Notice, to impose financial penalties on the provider concerned. The amount of the fine should not exceed 10 per cent of the turnover of the notified provider's 'relevant business'[28] for the period OFCOM considers to be appropriate and proportionate to the contravention (section 97(1)). In accordance with section 392 of the Communications Act, OFCOM has published Penalty Guidelines which state that when setting a penalty it is likely to consider as a starting point:

---

27   See, for example, *Notification of Contravention of General Condition 4.2—Notice served on Virgin Media Limited* (OFCOM, 21 December 2007).

28   Relevant business means, effectively, the provision of an ECN or an ECS and is defined at section 97(5) of the Communications Act.

- the seriousness of the contravention;

- any precedents set by previous cases; and

- the need to ensure the threat of penalties will act as a sufficient incentive to comply[29].

A person served with an Enforcement Notice will be under a statutory duty to comply; the Notice is enforceable in civil proceedings by OFCOM, whether by an injunction, specific performance (in Scotland) or any other appropriate relief (section 95(5) and (6)).

The process set out in the Communications Act under Sections 94–96 is curious. Following receipt of a Notification, the notified provider is entitled to make representations to OFCOM. OFCOM may subsequently decide to issue an Enforcement Notice for failure to comply with the Notification. The process does not therefore envisage that OFCOM will make any final determination that a contravention has actually occurred. Rather the recipient of an Enforcement Notice must immediately seek to remedy the consequences of their breach to avoid further steps being taken against them.

5.2 OFCOM's Enforcement Powers in Practice

On 23 June 2006, OFCOM issued an Enforcement Notice to Just Telecomms UK Limited (JTUK) concerning its contravention of General Condition 14.3(a), which requires compliance with the provisions of OFCOM's Code of Practice for Sales and Marketing. Specifically, OFCOM found (following investigations carried out in conjunction with Trading Standards officers) that JTUK had misled customers into thinking it was part of/affiliated to BT, failed to make clear to customers that they were entering into long-term contracts with punitive charges (of up to £395 per line) for early disconnection, and taken disproportionate action against customers who failed to pay those charges immediately. OFCOM had previously issued a Section 94 Notification to the company on 14 March 2006.

Although JTUK had, following the Notification taken certain steps to remedy the consequences of its breach, notably in writing to customers on appropriate terms, OFCOM was not satisfied these were sufficient (for example, the letters covered only a small proportion of all the affected customers). The Enforcement Notice set out several steps JTUK was required to take to ensure compliance, which included an obligation to write to all current and former customers from whom termination charges had been collected to bring the Enforcement Notice to their attention (and provide them with the means to seek redress). At the same time, in view of the limited steps undertaken by JTUK to remedy their breaches, OFCOM decided to issue a penalty set at the maximum level of 10 per cent of turnover. In so doing it was clearly intending to send a message to would be miscreants: this was the first occasion on which it had issued a penalty under Section 96[30].

---

29 *OFCOM Penalty Guidelines* (Section 392 of the Communications Act). Note that these guidelines apply to all penalties imposed by OFCOM under the Communications Act or any other enactment, other than the Competition Act 1998 (see Section 7.1.5 below).

30 *Enforcement and penalty notification given to Just Telecomms UK Limited*: 23 June 2006.

186

5.3 Powers to Suspend or Restrict Services

In serious or urgent cases involving a provider to whom OFCOM is entitled to issue a Notification, OFCOM may give a direction suspending or restricting that provider's right to provide an ECN or ECS or any associated facilities (sections 98(4) and 100). A provider who breaches a direction issued in this manner is guilty of a criminal offence. In view of the seriousness of these measures, OFCOM must first give notice of any proposed directions to the provider concerned and afford them a reasonable opportunity to make representations in their defence (sections 99 and 102)[31].

5.4 Third Party Damages Actions

Section 104 of the Communications Act makes separate provision for a third party who suffers damage following a breach of a condition under section 45 of the Act to bring civil proceedings against the infringing party, provided they first obtain OFCOM's consent to do so.

6 Dispute Resolution Powers

Article 20 of the Framework Directive imposes an obligation on EU Member States to confer powers on national regulatory authorities to resolve disputes between electronic communications providers arising in connection with any obligations under the EC Framework directives. Article 5(4) of the Access Directive reaffirms this obligation specifically in relation to disputes about access and interconnection. Section 185 of the Communications Act implements these obligations.

6.1 Definition of 'Dispute'

Section 185 defines a dispute as:

- one relating to the provision of network access if it is a dispute between different communications providers (section 185(1); or

- any other dispute, provided it is a dispute between different communications providers relating to rights or obligations conferred or imposed by or under Part 2 of the Communications Act or any enactment relating to the management of radio spectrum (section 185(2))[32].

In *Orange v OFCOM (BT Interconnection dispute)*,[33] the CAT was asked to look at whether OFCOM had been right to characterise negotiations between BT and Orange on the price BT was required to pay for mobile call termination services as a dispute within the meaning of section 185. Those negotiations arose out of provisions of the standard interconnection agreement (the SIA) which BT enters into with each of the MNOs which, among other things, sets out the charge for

---

31 OFCOM has not, to date, had occasion to exercise these powers.
32 Certain disputes are executed from the scope of section 185(2) by reference to section 185(7).
33 *Orange Personal Communications v OFCOM* [2007] CAT 36. Following the CAT's decision on these points, Orange subsequently withdrew the remainder of its appeal as to OFCOM's decision to accept jurisdiction.

termination services on each contracting party's network and the mechanism by which that charge can be varied.

Orange argued that OFCOM's powers under section 185 of the Communications Act should be narrowly interpreted to circumstances where an interconnection was yet to be established or was somehow at risk. This was rejected by the CAT. The CAT was particularly concerned that OFCOM's jurisdiction should not be dependent on its interpretation of the terms of the agreement at issue (and, in particular, any termination provisions). The CAT held the underlying purpose of the Access and Framework directives on which OFCOM's section 185 powers were based is to ensure that providers uphold their obligations on a lasting basis and not only at the outset of commercial negotiations or at the stage at which termination becomes a viable option.

6.2 Handling the dispute: OFCOM's Decision-making Powers

Once it has decided there is a dispute with the meaning of section 185 of the Communications Act, OFCOM must then determine whether it is appropriate for it to handle the dispute with regard to the grounds made out in section 186(3), which provides as follows:

'Unless they [OFCOM] consider

> *(a)   that there are alternative means available for resolving the dispute,*
>
> *(b)   that a resolution of the dispute by those means would be consistent with the Community requirements set out in Section 4, and*
>
> *(c)   that a prompt and satisfactory resolution of the dispute is likely if those alternative means are used for resolving it,*
>
> *their decision must be a decision that it is appropriate for them to handle the dispute.'*[34]

It is clear therefore that OFCOM has a limited degree of discretion as to whether or not to accept a dispute. Useful guidance on OFCOM's approach to this difficult issue is set out in the relevant Section of its Enforcement Guidelines. Those Guidelines emphasise that OFCOM cannot accept a dispute unless the applicant can provide a sufficient level of detail of the history of commercial negotiations, including the relevant contractual documentation and evidence of the failure of negotiations to resolve the dispute. It may be, for example, that the parties to the dispute are under a contractual obligation to attempt to negotiate a settlement—in such circumstances the evidential burden on them to persuade OFCOM to adjudicate will undoubtedly be higher.

Balanced against these considerations, the third of the factors set out in Section 186(3) makes clear that any alternative means suggested must be capable of achieving a 'prompt and satisfactory resolution'. Alternative forms of dispute resolution invariably rely on a willingness by all sides to co-operate. Consequently, unless each of the parties concerned agree on an alternative means as being

---

34   Recital 32 of the Framework Directive provides that

> 'an aggrieved party that has negotiated in good faith but failed to reach agreement should be able to call on the national regulatory authority to resolve the dispute'.

appropriate, it is unlikely that a resolution will be found. It can be seen therefore that in many cases OFCOM will have no option but to take jurisdiction where it finds that one or both of the section 185 grounds is made out.

Once it decides whether or not to take jurisdiction over a dispute, OFCOM will inform the parties of its decision. The date of this decision has procedural significance since, other than in exceptional circumstances, OFCOM must make its final determination (see below) within four months (section 188(5)). Alternatively, if OFCOM decides not to handle a dispute, but that dispute is not resolved by other means within a further four months of the date of its decision, either party to the dispute may refer it back to OFCOM (section 186(6)). In those circumstances, the parties would be expected to provide details of the steps they have taken to try to resolve their dispute by alternative means and details as to why these have been unsuccessful. Once it receives such information, OFCOM is duty bound to consider the dispute and reach a determination on resolving it within a further four months.

6.3 Reaching a Determination on the Dispute: Scope of OFCOM's Powers

Where it commences an investigation into a dispute, OFCOM will publish an entry in its Competition Bulletin setting out the scope of that investigation and, in particular, the issues it proposes to deal with. Before making a final determination, OFCOM will ordinarily invite comments from the parties concerned and, if it considers a wider consultation is necessary, publish details on its website seeking comments from third parties. In accordance with the requirements of the Communications Act (section 188(7)), a copy of the final determination and an explanatory statement of OFCOM's reasons will be sent to the parties. It is also normal practice for a non-confidential version of the same to go on OFCOM's website. Publication in this manner is also important in the context of the potential for a determination to be appealed under section 192 of the Communications Act (see Section 9.1 below).

In making a determination resolving a dispute, section 190 of the Communications Act provides that OFCOM may:

- make a declaration setting out the rights and obligations of the parties;

- fix the terms or conditions of transactions between the parties and oblige them to enter into a transaction on those terms; and

- require the payment of sums by one party to another by way of an adjustment of an over or under payment (where it has made a determination as to the amount of any charge).

These powers apply in relation to all disputes, other than those which relate to rights and obligations conferred or imposed by or under any provisions of the Communications Act which relate to the management of radio spectrum.[35]

The Communications Act does not give any specific guidance as to how OFCOM should exercise its section 190 powers in practice. However, in reviewing OFCOM's approach, case law in the CAT emphasises that there are essentially two tasks in

---

35 Section 190(3) provides that OFCOM's main power in such cases is to make a declaration setting out the rights and obligations of the parties to the dispute.

any determination by OFCOM: (a) to make an adjudication between the respective interests of the parties to the dispute; and (b) to ensure that its approach is consistent with its wider statutory duties under sections 3 and 4 of the Communications Act (see Sections 2.1 and 2.3 above). Further, when setting price or other contract terms, it is incumbent on OFCOM to ensure that it reaches a settlement which is 'reasonable' in all the circumstances. For example: *T-Mobile, BT, Hutchinson 3G, Cable & Wireless v OFCOM*)[36].

In this case, OFCOM had published two determinations concerning charges for mobile call termination (MCT) as between: (i) BT and each of the MNOs; and (ii) Hutchinson 3G (H3G) and each of 02 and Orange. Both determinations concerned blended charges (for 2G and 3G termination) charged or proposed by the MNOs. At the relevant time, mobile termination charges set by the 2G MNOs (ie excluding H3G) were subject to SMP conditions imposed following OFCOM's review of the wholesale MCT market[37]. OFCOM concluded in each of the determinations that the MNOs were able to set a blended rate which was higher than the regulated 2G charges. Appeals against each determination were combined by the CAT since they touched on similar issues.

The CAT found that OFCOM had wrongly relied on the previous exercise of its SMP powers to justify its approach to the disputes. Rather, OFCOM's dispute resolution powers should be approached as an autonomous regulatory process in respect of which specific regard must be had to its statutory duties. It was not permissible for OFCOM to argue that, since it had complied with its statutory duties in respect of its earlier SMP review, it could correspondingly rely on the results of that review to argue it had complied with those duties under a separate dispute resolution process.

The CAT was particularly critical of OFCOM's approach to the MCT charges set by BT where it had looked at whether, at the rate proposed, BT could make a profit from achieving connectivity with each MNO (the so-called 'gains from trade test'). In view of OFCOM's principal duty under section 3(1) (b) to further the interests of consumers, the CAT held that it was under a duty to assess the impact of the charges at issue on retail customers.

It is clear therefore that, where it is involved in adjudicating on what constitutes a reasonable price, OFCOM will be under a duty to undertake a detailed cost-based analysis in order to justify its conclusions. This is undoubtedly an onerous burden, particularly in view of the relatively short time frame (four months) within which determinations must be reached. The CAT recognised this point, but emphasised that, whilst OFCOM had a certain degree of discretion as to how to approach this task, it would at the very least expect some form of analysis as to the relationship between prices and costs in the final determination.

6.4 Impact of a Determination

A determination by OFCOM resolving a dispute binds all parties to it (section 190(8)). However, the Communications Act also recognises the possibility of parallel actions being undertaken, whether by the parties or OFCOM itself. In this

---

36 *T-Mobile, BT, Hutchinson 3G, Cable & Wireless and others v OFCOM* [2008] CAT 12.
37 See Chapter 3 above for an explanation of OFCOM's powers in relation to SMP conditions.

way, the referral of a dispute to OFCOM does not prevent any person (including one of the parties to the dispute) from starting court proceedings in respect of those matters under dispute (section 186(1)), nor OFCOM from exercising any of its other enforcement powers under the Act (section 186(2)). For example, OFCOM may in addition to considering a dispute use its powers to set, modify or revoke such general or specific conditions as it deems appropriate.

7 Concurrent Powers

It has been one of central government's stated policy objectives that, wherever possible, oversight of the telecommunications industry should be undertaken through the application of *ex post* competition law rather than *ex ante* regulation. A similar theme also pervades the European Commission's Framework Package, both in its current and proposed amended form. As has already been discussed above, the Communications Act imposes a specific duty on OFCOM to 'further the interests of consumers in relevant markets, where appropriate by promoting competition' (section 3(1)).

Within the UK, principal responsibility for the enforcement of competition law across all sectors of the economy lies with the Office of Fair Trading (OFT) in accordance with the powers granted to it under the Competition Act 1998 and the Enterprise Act 2002. However, OFCOM has (since 29 December 2003) exercised certain of these powers concurrently with the OFT, specifically in relation to 'activities connected with communications matters'[38].

7.1 Competition Act 1998

UK competition law was substantially reformed by the Competition Act 1998 which came into force on 1 March 2000. The Act contains two prohibitions; the so-called 'Chapter I' and 'Chapter II' prohibitions, contained respectively in sections 2 and 18 of the Competition Act 1998.

Under section 371 of the Communications Act, OFCOM was given concurrent power to enforce the Chapter I and Chapter II prohibitions. The exercise of OFCOM's concurrent powers are subject to a statutory instrument which seeks to ensure consistency of approach between the 'concurrent regulators'[39] and also deals with any instances where there is a dispute over who should take jurisdiction in a particular case[40].

A table showing all the decisions reached by OFCOM under the Competition Act 1998 to date is set out at the end of this Chapter.

---

38 Section 369 of the Communications Act defines communications matters for this purpose as being limited to the provision of electronic communications networks, services, facilities, as well as apparatus and broadcasting and related matters.
39 As well as OFCOM a number of other regulators are given contractual powers in relation to the industries they regulate.
40 SI 2004/1077. See also *Concurrent application to regulated industries*, OFT Guideline 405 and the recent decision of the High Court in *R v Office of Fair Trading* [2009] EWHC57 which looked at the interplay between OFCOM and the OFT.

7.1.1   The Competition Rules: Introduction

Chapter I and Chapter II of the Competition Act 1998 are closely modelled on the equivalent prohibitions contained in Articles 81 and 82 of the EC Treaty. Articles 81 and 82 are only applicable where an appreciable effect on trade between Member States can be made out; this threshold sets out the boundary between the application of Community and national competition law. However, in practice this distinction is unimportant given the similarity between the UK provisions and their EC counterparts.[41] Article 5 of Regulation 1/2003[42] (the Modernisation Regulation) provides that national competition authorities should have the power to apply Articles 81 and 82 in individual cases. OFCOM has also been designated for this purpose.

A fuller analysis of the competition rules can be found elsewhere[43], but for present purposes it is relevant to give a brief overview.

7.1.2   Prohibition on Anti-competitive Agreements

Chapter I of the Competition Act 1998 and Article 81 each prohibit agreements, decisions by associations of undertakings[44] and concerted practices which have as their object or effect the restriction of competition, unless they can benefit from an exemption.

There are a number of important features of the prohibition. First, it is not just written contracts which are caught; a 'concerted practice' can be any form of practical co-operation knowingly entered into by two or more parties which is intended to substitute such co-operation for the normal conditions of competition. Second, certain agreements will, by their very nature, be found to have an anti-competitive 'object' in which case they automatically breach the prohibition. These are known as 'hardcore infringements' and include, for example, price fixing and limiting output. Alternatively, where no such object can be found, the lawfulness of an agreement will be tested according to its anti-competitive effects (based on the impact of the agreement on the relevant market).

Each of the EC and UK prohibitions provide a mechanism for anti-competitive agreements to be 'exempted' provided they fulfil certain criteria which, in broad terms, ensure that the economic and social benefits of the agreement outweigh its anti-competitive impact[45].

---

41   The Competition Act 1998 includes a specific provision (section 60) ensuring that 'as far as possible' questions arising under the competition rules in the UK are dealt with in a manner which is consistent with the treatment of corresponding questions under EC law.

42   OJ [2003] L1/1.

43   See for example Whish *Competition Law* (OUP, 6th ed 2008).

44   The term 'undertaking' for the purposes of the application of the competition rules is not defined anywhere in either the EC Treaty or UK legislation. However, there is a large body of case law on this issue which, broadly speaking, provides that an undertaking is an entity which is engaged in an economic activity.

45   The exemption criteria are set out in section 9 of the Competition Act 1998 and Article 81(3) of the EC Treaty. Previously, the parties to an agreement could 'notify' it to the competition authorities asking for a ruling that it did not infringe Article 81/section 2 or that, in the alternative, it satisfied the conditions for exemption. However, the Modernisation Regulation ended this practice and it is now up to the parties themselves to assess the compatibility of their agreement which the competition rules; an agreement is deemed to be lawful until such time as a competition authority or court declares otherwise.

In practice, there have been relatively few cases involving anti-competitive agreements between telecoms operators: all but one of the decisions issued by OFCOM to date using its competition law powers has exclusively concerned Chapter II/Article 82 issues. The one exception was a case involving the provision of 'media access services' (for those with hearing or sight disabilities) to TV programmes, in which OFCOM considered both Chapter I/Article 81 and Chapter II/Article 82 issues as regards the length of an exclusive contract between BBC Broadcast and Channel 4[46]. However, this is not to say that the Chapter I/Article 81 provisions are not relevant to this sector. For example, the European Commission has previously taken decisions under Article 81 concerning agreements by mobile operators as to network sharing[47].

7.1.3 Prohibition of Abuse of a Dominant Position

Whereas Chapter I/Article 81 are concerned with agreements, decisions or concerted practices between undertakings, Chapter II of the Competition Act 1998[48] and Article 82 of the EC Treaty concern unilateral conduct and prohibit any abuse by an undertaking of its dominant position.

Chapter II/Article 82 apply only where one undertaking has a dominant position or two or more undertakings are collectively dominant[49]. There are two stages to undertaking an analysis of dominance: the definition of the 'relevant market', and an analysis of the relative economic strength of the undertaking in that market.

For competition law purposes, the relevant market consists of the product market (comprising those products which are substitutable for one another) and the geographic market (the geographic area within which the conditions for competition are homogenous). Within the EU Framework Package, the Commission has published a 'recommendation' on the definition relevant markets in the electronic communications sector (Recommendation) to which national regulators are under a duty to have the utmost regard in accordance with article 15 of the Framework Directive[50]. Whilst this document may be persuasive for competition law enforcement purposes, it is also important to recognise that the criteria for determining the relevant market will depend on the objective being pursued by the relevant legislation under which the assessment is being made. Whereas regulation (such as the SMP provisions takes place on a forward-looking (*ex ante*) basis, competition law enforcement is *ex post* and necessarily takes into account evidence based on events which have already happened in a market. This point is made explicitly in the Recommendation[51]. It is therefore perfectly possible that a relevant market defined by OFCOM for Chapter II/Article 82 purposes will be different to that used for the purposes of OFCOM's regulatory powers.

---

46   *BBC Broadcast: provision of media access services* (CW/00842/06/05)—see table of Competition Act decisions below.

47   See for example, *T-Mobile Deutschland/02 Germany: Network sharing* OJ [2004] L75/32. This decision was later overturned on appeal by the Court of First Instance.

48   Section 18 of the Competition Act 1998.

49   Collective dominance is a controversial topic and has been the subject of much debate and jurisprudence, especially in the field of merger control, but basically reflects the idea that two or more companies who are not individually dominant may together be collectively dominant on account of economic or contractual links between them.

50   *Commission Recommendation on relevant product and service markets within the electronic communications sector (Second edition)*, OJ [2007] L344/65.

51   Ibid, at Recital 16.

Specific guidance as to market definition for competition law purposes is provided by the European Commissions Notice on the Definition of the Relevant Market[52] and within the OFT's guidelines on the application of the Competition Act 1998 in the telecommunications sector[53].

The European Court of Justice defines dominance as:

'a position of economic strength enjoyed by an undertaking on a market which enables it to prevent effective competition being maintained on the relevant market by affording it the power to act to an appreciable extent independently of its competitors, its customers and ultimately of the consumers.'[54]

The basic idea is therefore that a company is in a position of market power which enables them to charge higher prices than if faced with effective competition, or otherwise exclude or deter competition from the market.

In assessing whether a company has such market power, a proxy which is often used is to look at market share. As a rule of thumb, a market share in excess of 50 per cent creates a rebuttable presumption of dominance[55], whereas an undertaking with a share of less than 40 per cent will not usually be dominant[56]. However, market share is only the starting point and it is necessary in any case to consider carefully the constraints to which an undertaking may be subject either from existing or potential competitors and/or customers.

Section 18(2) and Article 82 each contain examples of abuse in identical terms, although it is an important principle of the application of the law on abusive behaviour that there is no exhaustive list of what constitutes an abuse. Within the telecommunications sector, amongst the most relevant categories of abuse are:

Predatory pricing: selling at below the marginal cost of production and thereby deliberately incurring losses in the short term with the effect of eliminating an existing competitor from the market and later being able to charge higher prices and recoup those losses;

Margin squeezing: where a firm is dominant in an upstream market and supplies a key input/service to undertakings (eg access to a network) with whom it also competes in a downstream market (eg retail supply of network services). A margin squeeze occurs where the firm seeks to make it difficult for its downstream rivals to compete by imposing a wholesale price for the key input which, if they charge the same price as the dominant company in the downstream market, makes them unprofitable;

Discriminatory pricing: charging customers in comparable positions different prices in circumstances where the costs of supply do not differ;

---

52 OJ [1997] C372/5.
53 OFT 417 (The Application of the Competition Act 1998 in the Telecommunications Sector). Note these guidelines preceded the Communications Act, hence the reference therein to the Director-General for Telecommunications rather than OFCOM.
54 *United Brands v Commission* Case 27/76 [1978] ECR 207.
55 *Case 53/85 Akzo BV v Commission* [1986] ECR 1965.
56 Exceptionally, in *British Airways v Commission* 2007 [ECR] I-2331, BA was found to be dominant with a market share of 39.7 per cent .

Excessive pricing: pricing which is excessive by reference to the economic value of the product or service provided;

Refusal to supply: refusing to supply a product or service, whether outright or because the terms on which that product or service are offered are such that the customer cannot reasonably be expected to purchase them (so-called 'constructive refusal'); and

Bundling/tying: where an operator ties the supply of products in a market in which it is dominant to the supply of products/services that are (or at least could be) supplied competitively on their own, eg where access to a network is only provided on condition that the customer also purchases certain equipment from the same supplier.

There is no explicit exemption mechanism under the Chapter II prohibition or Article 82 in the same way as under Chapter I/Article 81. However, a dominant undertaking can raise a defence to an alleged abuse where it can show that it has some form of 'objective justification' for its behaviour.

The issue of objective justification was central in the *Floe Telecom* case where OFCOM found that Vodafone had not infringed Chapter II or Article 82 by disconnecting services provided to Floe for use in GSM gateways (ie a refusal to supply) because Floe was requiring Vodafone to act illegally in violation of the Wireless Telegraphy Act 1949. On appeal, the CAT accepted that if Floe's behaviour had been clearly illegal this would have been sufficient grounds for Vodafone to refuse to supply, but found OFCOM's analysis of the legality or otherwise of Floe's request to be inadequate[57]. A second decision by OFCOM in the case reaching the same conclusion which was also appealed by Floe, was ultimately decided by the CAT on different grounds (see further at Section 9.5 below)[58].

The same considerations were also applied by OFCOM in its *Re-investigation of a complaint from VIP Communications Ltd* concerning a refusal by T-Mobile to supply services[59].

7.1.4   The Application of OFCOM's Competition Act Powers

The point has already been made above that all but one of the decisions made by OFCOM using its Competition Act powers have exclusively concerned the application of the Chapter II prohibition or Article 82 of the EC Treaty. What is also significant is that OFCOM has not, to date, yet reached a decision finding that a company has infringed the competition rules. This is, at least in part, a reflection of the complexities involved in bringing such cases, notably those involving pricing practices.

---

57   Case No 1024/2/3/04 [2004] CAT 18.
58   Case 1024/2/3/04 2004 [CAT] 18 and [2006] CAT 17. Note that aspects of the CAT's second decision were subsequently appealed to the Court of Appeal see *Office of Communications & ANR v Floe Telecom* [2009] ENCA Civ 47.
59   OFCOM's decision in that case is also on appeal to the CAT (see Case 1027/2/3/04 *VIP Communications Ltd (in administration v OFCOM)*, the appeal was stayed pending final resolution of the *Floe Telecom* case and, at the time of writing, is yet to be finally resolved.

These difficulties are perhaps best illustrated in the number of completed investigations by OFCOM into allegations of margin squeezing[60]. In each case, OFCOM found no evidence of a margin squeeze or, even if there was a margin squeeze, no evidence that it would cause customers of the dominant companies to pay more than they would otherwise have paid in a competitive market. In an ongoing investigation into whether BT operated an unlawful margin squeeze through the pricing of its residential broadband services, which effectively began in November 2003, OFCOM has issued three separate statement of objections to BT setting out its intention to reach a decision finding that it has infringed Chapter II and Article 82, but on the basis of the representations received in response it has yet to reach a final decision[61].

7.1.5   Enforcement

A 'live' issue in the enforcement of competition law is the extent to which competition regulators should be able to prioritise their casework. Both the EC and UK competition rules are directly enforceable by affected third parties before the courts by way of an action for damages or injunction. In this way, there has been a move amongst the regulators to encourage parties to air their commercial grievances before the courts, thereby leaving those regulators scope to focus their resources on the most important cases[62].

Practical Issue: Direct (private) actions

Historically within Europe, public enforcement of competition law has taken on a much more important role than private enforcement through the courts/ tribunals constituted for that purpose. This contrasts markedly with the US where private actions tend to form the large majority of competition law (antitrust) cases, encouraged in particular by the 'class action' culture which pervades that legal system.

Competition authorities/regulators are increasingly eager to promote the use by companies and individuals of private court-based enforcement (direct actions) to enforce their legal rights. This is reflected in those bodies opting not to proceed with investigations into complaints brought to their attention unless it is felt that such an investigation would be a worthwhile use of their limited resources. As alluded to above (see Section 7.1.5)), OFCOM screens carefully all complaints and carries out an initial evaluation as to the merits or otherwise of opening a formal investigation based on factors, including a balancing of the likely internal resources needed to conduct such an investigation to its conclusion as against the risks to the interests of consumers as a result of the alleged infringing behaviour (see OFCOM's Enforcement Guidelines at paragraphs 4.42–4.47).

Examples of successful direct actions do exist. For example, in July 2007, Software Cellular Network Limited (trading as Truphone) obtained an injunction against T-Mobile requiring it to connect calls to Truphone numbers. T-Mobile had been

---

60   *Alleged anti-competitive practices by BT in relation to BT Openworld's consumer broadband products, Investigation against BT about potential anti-competitive behaviour, BT 0845 and 0870 retail price change* and *Suspected Margin Squeeze by Vodafone, O2, Orange and T-Mobile* (see table above).
61   This investigation was precipitated by a complaint issued by Freeserve (now part of Orange UK) in March 2002.
62   See also Section 9.3 below.

refusing to do so on the basis that the termination fees demanded by Truphone were too high, behaviour Truphone argued amounted to an unlawful refusal to supply contrary to Chapter II of the Competition Act 1998 (see Section 7.1 above).[63]

Many cases are also likely not to be in the public knowledge as they settle in advance of reaching the courts. In reality, however, direct action remains a relatively underused tool. There are a number of factors which explain this, the most important of which is the significant cost in pursuing any litigation through the UK courts. This is particularly relevant in competition law cases where, without a decision by a regulator on which to rely,[64] fundamental issues such as market definition require detailed evidence and analysis before the claimant has even got into the alleged infringement(s).

Action is being taken at both European and UK level to address these issues. A study commissioned by the European Commission which was published in August 2004 (the 'Ashursts report') revealed that across the EU there were numerous obstacles to successful private actions for damages for competition law infringement which had led to a 'total underdevelopment' in such cases.[65] This study led the Commission to publish two (Green and White) papers setting out its proposals for facilitating more actions.[66] In a similar vein, the OFT has issued recommendations to Government setting out what changes could be made to the UK litigation rules to alleviate the situation.[67] It remains to be seen whether these initiatives have the desired effect.

7.1.6 Procedure

OFCOM may carry out an investigation under the Competition Act 1998 if there are 'reasonable grounds for suspecting' that either of the Chapter I or II prohibitions in the Competition Act 1998 or Articles 81 or 82 have been infringed (section 25). This power is entirely discretionary: OFCOM does not have any duty to investigate. However, where it does so, its information gathering powers under the Competition Act 1998 are considerable and enable OFCOM (under appropriate circumstances) to:

- require a person to produce specified documents or specified information which it considers to be relevant to its investigation (section 26);

- enter business premises without a warrant (section 27); and

- enter and search any premises (either business or domestic) with a warrant (sections 28 and 28A).

---

63 *Software Cellular v T-Mobile (UK) Limited* (17 July 2007, Knowles DHJ).
64 Under Section 47A of the Competition Act 1998, a decision that Chapter I/Article 81 or Chapter II/Article 82 has been infringed (and which hasn't been successfully appealed) is binding in any subsequent ('follow-on') action for damages which can be brought directly to the CAT. Section 58 of the Competition Act 1998 provides for the same principle in respect of follow-on actions brought in the High Court.
65 Study on the conditions of claims for damages in the case of infringement of EC competition rules, August 2004.
66 COM (2005) 672 final (Green Paper), COM (2008) 154 final (White Paper).
67 *Private actions in competition law: effective redress for consumers and business.* OFT 916 resp.

Having conducted an investigation pursuant to these powers, OFCOM may decide not to proceed any further. This can occur for a variety of reasons including administrative priority (ie on account of the considerations referred to above), insufficient evidence to establish an infringement, or other grounds unrelated to the merits of the case (eg withdrawal of a complaint). Another alternative is that OFCOM accepts binding commitments from the party under investigation which are sufficient to address its competition concerns[68]. Whatever the grounds on which it decides to close a case, it is clear that OFCOM will need to exercise great care in view of the rights of appeal of parties affected by any OFCOM 'decision' (see Section 9 below).

Where it finds there are sufficient grounds to establish an infringement, OFCOM will issue a statement of objections (SO) to the party concerned which sets out the case against them and the action it proposes to take, as well as giving that party a period of time in which to respond to the allegations being made against them[69]. Interested third parties, particularly any complainants, are also likely to be given an opportunity to make representations on the SO.

7.1.7   Interim Measures

A complainant who has persuaded OFCOM to conduct an investigation under Section 25 may not be able to wait for the outcome of that investigation to continue as a viable commercial entity. Section 35 of the Competition Act 1998 allows OFCOM to adopt interim measures, if necessary as a matter of urgency to prevent serious irreparable damage to a particular person(s) (eg an order to re-start supplies of goods or a service that has been withdrawn by a dominant supplier). OFCOM must give notice to the affected persons before making any such directions to give them a chance to make representations in response. In effect, this process operates in the same way as that concerning an application for an interim injunction in the High Court (see *Software Cellular Network* case cited above).

7.1.8   Directions

If it proceeds to a final infringement decision, OFCOM has powers to give directions bringing the infringement to an end (Competition Act 1998, sections 32 and 33) and to fine the party concerned up to the statutory maximum of 10 per cent of its worldwide turnover in the business year preceding the decision (section 36)[70]. Clearly, this can be a substantial amount.

7.2   Market Investigations Under the Enterprise Act 2002

In addition to its concurrent powers under the Competition Act 1998, section 370 of the Communications Act confers certain concurrent powers on OFCOM

---

68   See Competition Act 1998, section 31A.
69   The recipient of a statement of objections also has other procedural rights, including the opportunity to access those documents held by OFCOM on its 'file' in relation to the investigation and to make oral representations. See further Competition Act 1998 (OFT Rules) Order 2004, SI 2004/1517, at Section 5.
70   See Competition Act 1998 (Determination of Turnover for Penalties) Order 2000, SI 2000/309, as amended by SI 2004/1259.

contained in the Enterprise Act 2002, most notably the ability to make 'market investigation references' to the Competition Commission. This power applies (in accordance with section 131 of the Enterprise Act 2002) where OFCOM has reasonable grounds for suspecting any 'feature or combination of features' of a market in the UK[71] prevent, restrict or distort competition. As in relation to its powers under the Competition Act 1998, section 369 of the Communications Act makes clear that OFCOM's market investigation powers under the Enterprise Act 2002 are limited to 'communications matters'.

On a market investigation reference from OFCOM, the Competition Commission (CC) will conduct a detailed investigation of the competitive workings of the affected market(s) (which can last up to two years) at the end of which, if it finds an adverse effect on competition, the CC will decide on suitable remedies to be applied to companies active in that/those market(s). Remedies can include, in appropriate circumstances, structural remedies (eg divestment) or (behavioural) undertakings by the companies concerned as to how they will deal with their competitors/customers in the future. However, provision is also made in the Enterprise Act 2002 for the regulator concerned to accept legally binding undertakings in lieu of (ie instead of) a reference to the CC (section 154(2)). It was these powers which OFCOM used in June 2005 when it accepted a comprehensive package of undertakings from BT (see Chapter 5 above) to avoid a CC investigation.

In accordance with section 162 of the Enterprise Act 2002, OFCOM is required to keep any enforcement undertakings under review and to ensure they are complied with. It is also required to consider whether, in view of a change of circumstances, there is a case for release, variation, or revocation of any one or more undertakings. BT has used this provision to seek its release from or variation of several of the original undertakings given to OFCOM in 2005[72].

OFCOM also has a role in considering any 'super-complaints' made under the Enterprise Act 2002 in relation to communications matters. Super complaints are made by consumer representative bodies designated by the Secretary of State for this purpose who believe that one or more features of a market may be harming consumers to a significant extent[73]. Upon receipt of a complaint, OFCOM must, within 90 days, publish a 'fast track' report on what action, if any, it intends to take (Enterprise Act 2002, section 11(2)). Such further action can include a more lengthy market study, enforcement action or a referral of the market to the CC in the same way as described above[74].

---

71 For a definition of what constitutes a 'feature' of a market for these purposes, see Enterprise Act 2002, section 131(2).

72 See, for example, the variation agreed between BT and OFCOM in removing BT from the obligation to provide IPStream and its successor products in all parts of the UK. Statement by OFCOM dated 4 December 2008.

73 See the Enterprise Act (Bodies Designated to make Super-complaints) (Amendment) Order 2005, SI 2005/2340.

74 In June 2008, OFCOM received a super complaint from the National Consumer Council relating to the cost of calls made by prisoners in England, Wales and Scotland. On 22 September 2008, OFCOM published its response in which it suggested that the issues identified by the complaint could most appropriately be dealt with by the government bodies responsible for the procurement of telephone services in prisons.

7.3 Sections 316–317 Communications Act: OFCOM's Sectoral Competition Powers

Sections 316–317 of the Communications Act empowers OFCOM to impose conditions on providers of (TV and radio) services licensed under the Broadcasting Acts which it considers '*appropriate for ensuring fair and effective competition*'. These are often referred to as OFCOM's 'sectoral competition powers'.

Section 316 provides specifically for OFCOM to include conditions (whether they take the form of compliance with a particular code or directions for that purpose) such as it considers appropriate for ensuring the licence holder does not:

'(a) *enter into or maintain any arrangements, or*

(b) *engage in any practice, which OFCOM considers, or would consider, to be prejudicial to fair and effective competition in the provision of licensed services or of connected services*'[75].

These powers were used by OFCOM in July 2006 to impose a new Code of Practice on TV broadcasters in respect of the cross-promotion of programmes, channels and other broadcast-related services; all TV broadcast licences include a licence condition which require the licence holder to comply with any code or guidance approved by OFCOM for the purpose of ensuring fair and effective competition[76].

Section 317 is broader and allows OFCOM to exercise a Broadcasting Act power for a 'competition purpose'. The Act defines a competition purpose by reference to the same test as that in italics above[77]. There is an obvious cross-over between the exercise of these powers and OFCOM's concurrent powers under the Competition Act 1998 (ie those set out at Section 7.1 above). To that end, before using its sectoral competition powers, OFCOM is under a duty to consider whether a more appropriate way of proceeding is under the Competition Act 1998[78].

The use of Sections 316/317 has recently attracted significant debate in the context of OFCOM's Pay-TV market investigation and, more specifically, its proposal to require Sky to make 'premium' (live Premier League football and first-run Hollywood films) content more widely available on a wholesale basis to other retailers. In making this proposal, OFCOM has concluded that the imposition of licence conditions to that effect would be a more appropriate way to proceed than to use those powers available to it under the Competition Act 1998 to remedy its concerns that Sky has market power in the supply of premium content and an incentive to limit access to its rivals in a manner which favours its own satellite platform[79]. In response, Sky has stated that it believes OFCOM would be applying its powers in an inappropriate manner and that any concerns as to the exercise of its market power could readily be dealt with under the Competition Act 1998.

Any decision taken pursuant to OFCOM's Section 316/317 powers is subject to appeal to the Competition Appeal Tribunal in the same way as any other 'decision'

75 Section 316(2) and (3).
76 See http://www.ofcom.org.uk/tv/ifi/codes/bcode/crosspromo/. This Code replaced rules which had been issued by OFCOM's predecessor the ITC in January 2002.
77 Communications Act, Section 317(9).
78 Communications Act, Section 317(2).
79 See http://www.ofcom.org.uk/consult/condocs/second_paytv/.

taken by OFCOM or the Secretary of State within the meaning given to that term under section 192 of the Communications Act (see Section 9 below).

7.4 Merger Control[80]

Certain corporate transactions fall to be examined by the UK competition authorities under the Enterprise Act 2002[81]. In most cases where a transaction qualifies for investigation it will be the OFT which, at least in the first instance, determines whether the transaction may proceed. The one exception is cases that involve 'public interest' considerations where it is the Secretary of State (advised by the regulatory authorities) who has the power of decision-making. Specifically its role is to decide whether to clear the transaction conditionally or unconditionally, or alternatively to 'refer' it to the Competition Commission for further assessment on the basis of any competition concerns which may arise as a result of the transaction.

Given its specialist knowledge in the area, where a transaction involves 'communications markets', OFCOM will routinely be consulted by the OFT for its views[82]. Further, in cases which involve possible issues of 'plurality of media ownership', OFCOM has a statutory role in advising the Secretary of State on what action, if any, he should take to protect the public interest[83].

8   Consumer Protection

As has been noted above, section 3(1) defines OFCOM's principal duties in carrying out its functions by reference to the protection of citizen and consumers' interests. Within this context, OFCOM takes extremely seriously its powers of consumer protection.

The existing legal framework for consumer protection in the communications sector is based on a complex set of regulations, some of which are sector specific and others which derive from broader cross-sector statutory regulations.

8.1 Sector-specific Regulations

8.1.1 *Enforcement of General Conditions*

It has already been seen that OFCOM seeks to protect consumers through the imposition of appropriate general conditions on communications providers, most notably those under general condition 14 which requires compliance with

---

80 For more detail on the UK merger control rules see *UK Merger Control: Law and Practice* (Sweet & Maxwell—2nd Edn).
81 To qualify for assessment, a transaction must involve: two or more enterprises ceasing to be distinct, and either: (i) that as a result of the transaction the combined enterprise supplies/acquires 25 per cent or more of any goods or services in the UK; or (ii) the turn over in the UK of the enterprise being taken over exceeds £70 million. This jurisdictional test is not always easy to apply in practice.
82 See, for example, the OFT's decision *Tiscali/Pipex* dated 17 August 2007 at paragraph 50.
83 These issues came to the force in respect of BSkyB's purchase of a (17.9 per cent) stake in ITV which was subsequently investigated by the competition authorities and the Secretary of State under the merger rules.

a number of codes of practice and dispute resolution procedures (see Chapter 3 above). OFCOM frequently uses its powers to monitor compliance with these general conditions. For example, it has looked at whether communications providers are complying with their obligations under general condition 14.7 to join a dispute resolution service and also taken a number of steps to deal with increasing complaints as to mis-selling by operators. A new general condition (GC23) on sales and marketing services was introduced in March 2009 which requires mobile service providers to stamp out mis-selling practices, in particular by ensuring all customers are authorised to enter a contract and are given all appropriate information at the point of sale.

8.1.2 Improper use of a Network or a Service

Section 127(1) criminalises the sending of offensive, indecent, obscene or menacing messages. Under section 127(2) a person will also be guilty of an offence if they send a message through a public electronic communications network (PECN) which is intended to cause annoyance, inconvenience or needless anxiety and which they know to be false or where they consistently make use of the PECN for that purpose. Punishment can include imprisonment and/or a fine.

8.1.3 Misuse of a Network or Service

If OFCOM reasonably suspects that a supplier is persistently misusing a network or service by causing persons to suffer annoyance, inconvenience or anxiety or if they have been reckless in doing so, it can launch an investigation. Where OFCOM determines there are reasonable grounds for believing an infringement has taken place, it may issue an enforcement notice requiring an end to the misuse. If the misuse continues, OFCOM can also impose a penalty of up to £50,000 per contravention (Communications Act, sections 128–131)[84].

OFCOM has used these powers frequently, especially in relation to what it sees as a growing problem of silent and abandoned calls generated by the automated calling systems engaged by companies[85].

8.1.4 Emergency Powers (section 132)

In circumstances where the Secretary of State has reasonable grounds for believing it is necessary to do so, they may direct OFCOM to take steps to suspend or otherwise restrict an electronic communications network or service provider (or a provider of associated facilities). Such action must be justified to protect the public from any threat to public safety or health, or in the interests of national security.

---

84 The maximum penalty was raised from £5,000 to £50,000 by the Secretary of State on 6 April 2006 (see SI/2006/1032).

85 In September 2008, Barclaycard were fined the maximum of £50,000 for breaching OFCOM's rules on silent and abandoned calls.

8.2 Rights and Obligations Deriving from Cross-sectoral Regulations

OFCOM has been appointed as a 'designated enforcer' under Part 8 of the Enterprise Act 2002[86] to take action to deal with matters such as doorstep selling, distance selling, unfair contract terms, consumer credit, TV advertising, sales of consumer goods and associated guarantees, data protection and privacy. More specifically, these powers enable OFCOM to enforce certain consumer protection legislation in respect of 'communications matters', including:

Unfair Terms in Consumer Contracts Regulations (UTCCRs)[87]: which provide that a contractual term that has not been individually negotiated in a consumer contract is unfair (and hence non-binding on the consumer) if, contrary to the requirement of good faith, it causes a significant imbalance in the rights and obligations of the parties to the detriment of the consumer.

Distance Selling Regulations[88]: which apply to contracts for goods or services to be supplied to a consumer where the contract is made exclusively by means of distance communication. This gives OFCOM the power to protect the rights of consumers where contracts are made, for example, via email, the telephone, letter or catalogue.

Consumer Protection from Unfair Trading Regulations 2008[89]: which came into force on 26 May 2008 and introduced a general prohibition on traders from engaging in unfair commercial (mainly marketing and selling) practices against consumers. The Regulations specifically ban misleading and aggressive commercial practices[90].

8.2.1 Enforcement

In practice, the enforcement process under Part 8 of the Enterprise Act will begin with a consumer complaint as to the conduct or practice of a particular supplier which will only be investigated by OFCOM where it considers that: (i) it is the regulator best placed to deal with the case; and (ii) if the behaviour is likely to harm the collective interests of consumers.

Where OFCOM believes that an infringement is taking place (and that it is best placed to act), it can agree undertakings with the party(ies) concerned as to their future business practices[91]. Undertakings given in this way are legally binding and enforceable by court order. Alternatively, in more serious cases, OFCOM may make an application to a court for an enforcement order designed to immediately stop the infringing behaviour.

---

86 As listed in the Enterprise Act 2002 (Part 8 Designated Enforcers: Criteria for Designation, Designation of Public Bodies as Designated Enforcers and Transitional Provisions) Order 2003 (SI 2003/1399).
87 SI 1999/2083.
88 SI 2000/2334.
89 SI 2008/1277. These Regulations, which implement the Unfair Commercial Practices Directive (Directive 2005/29/EC), also replace a wide range of existing consumer legislation for example much of the Trade Descriptions Act 1968.
90 A commercial practice is unfair if it contravenes the requirements of professional diligence, and materially distorts the economic behaviour of the 'average' consumer in relation to a product (or is likely to do so). A list of 31 practices which are considered unfair are included in the Regulations.
91 See for example, the undertakings agreed as between Phones 4U and OFCOM concerning the sale of mobile handsets and contracts 'Phones 4U probe', press release dated 10 November 2008.

Specifically in relation to UTCCRs, if a provider does not amend its terms and OFCOM's concerns are not alleviated, it may take the provider to court and seek a ruling on the legitimacy of those terms. Only the court can ultimately decide whether a term is unfair or not.

9 Appeals to the Competition Appeals Tribunal (CAT)

The CAT as it operates today was established under the Enterprise Act 2002[92], although it first came into being as the Competition Commission Appeals Tribunal under the Competition Act 1998[93]. Otherwise than in respect of price control matters (see Section 9.6 below), appeals in the CAT are governed by the Competition Appeal Tribunal Rules 2003 (Tribunal Rules 2003)[94].

The CAT has published a useful practical guide as to its procedural rules.[95] More specifically, the guide sets out the general principles which the CAT applies to all appeals in order to meet its overriding objective to ensure the parties concerned are on an equal footing, that unnecessary expense is saved and that appeals are dealt with expeditiously and fairly[96].

Despite having been in existence only a relatively short period of time, the CAT has become a fundamental part of the telecoms enforcement regime in the UK, not least because of the powers entrusted on it to hear appeals against 'decisions' taken by OFCOM or the Secretary of State under the Communications Act. Indeed, given the complexities faced by OFCOM in the exercise of its regulatory powers, it is perhaps no surprise that a high proportion of all the cases dealt with by the CAT since its inception have involved the telecommunications industry.

9.1 Rights of Appeal

Section 192 of the Communications Act specifies those decisions of OFCOM or the Secretary of State which can be appealed to the CAT. In relation to OFCOM, the rights of appeal apply to any decision which it takes pursuant to the exercise of its powers to regulate electronic communications networks and services under Part 2 of the Communications Act, other than those listed in Schedule 8 to the Act which may not be appealed against.

The right of appeal under Section 192 also applies to a decision by OFCOM made under the Wireless Telegraphy Act 2006 (see Chapter 7 above), that is not a decision specified in Schedule 8 to the Communications Act. The CAT recently had occasion to look at this area in an appeal by T-Mobile and O2 against the way in which OFCOM had conducted an auction of two bands of radio spectrum. The CAT concluded that the disputed 'decisions' were covered by the exceptions in Schedule 8 as being decisions made pursuant to section 14 of the Wireless Telegraphy Act and, as such, any right of appeal lay in judicial review proceedings before the High

---

92 Enterprise Act 2002, Section 12 and Schedule 2.
93 Competition Act 1998, Section 45(7) and Schedule 7.
94 SI 2003/1372, as amended by the Competition Appeal Tribunal (Amendment and Communications Act Appeals) Rules 2004, SI 2004/2068.
95 A Guide to Proceedings, October 2005. Available at www.catribunal.org.uk.
96 See Guide to Proceedings, paragraph 3.4.

Court[97]. It also rejected an argument put forward by the appellants that Article 4 of the Framework Directive, (which provides for Member States to ensure that effective mechanisms exist at national level for rights of appeal by communications providers against decisions by regulatory authorities), necessitated a full appeal on the merits of OFCOM's decisions. The Court of Appeal subsequently upheld this decision[98].

Decisions undertaken by OFCOM pursuant to its concurrent competition powers to apply the Competition Act 1998 (see Section 7.1 above) are also appealable to the CAT provided they come within the definition of an 'appealable decision' set out in Section 46 of that Act. Similarly, an appeal can be brought against a decision taken by OFCOM in respect of its market investigation powers under the Enterprise Act 2002.

Section 192(1)(d) of the Communications Act lists those decisions of the Secretary of State which are subject to appeal before the CAT[99].

9.2 Appealable Decisions Under the Competition Act

As detailed in 9.1 above, section 46 of the Competition Act 1998 specifies those decisions which can be appealed to the CAT. In particular, reference is made in the Act to any decision as to whether Chapter I or II of the Competition Act or their EC Treaty equivalents 'has been infringed'. Difficulties have arisen from this formulation. In particular, where a regulator (eg OFCOM) decides to reject a complaint, it has been argued that this constitutes an implicit finding of 'non-infringement' of the relevant prohibition which should therefore be subject to appeal.

The CAT's approach to this issue is to look at substance not form; a regulator cannot escape the possibility of an appeal merely by stating it has not reached any decision on a complaint. In *Freeserve Plc v Director-General of Telecommunications (DGFT)*[100], OFCOM's predecessor as telecoms regulator, the DGFT, had, following a preliminary investigation, rejected a complaint that BT was abusing its dominant position in relation to the provision of broadband services on the basis that he had insufficient information (evidence) to prove any anti-competitive conduct. In a preliminary judgment following an appeal by Freeserve, the CAT held that the DGFT's rejection letter to Freeserve was an appealable non-infringement decision. In so doing, it placed significant weight on the fact that the DGFT had in his letter to Freeserve addressed in fairly full terms each aspect of Freeserve's original complaint.

Since *Freeserve* and a number of other cases where the CAT found that the rejection of a competition law complaint constituted an appealable decision[101], due

97 T-*Mobile (UK) Ltd v Office of Communications* and *Telefonica 02 Limited v Office of Communications* [2008] CAT 15.
98 T-*Mobile (UK) Ltd & Anor v OFCOM* [2008] EWCA Civ 1373.
99 Namely: (i) a direction in respect of OFCOM's network and spectrum functions (pursuant to section 5 of the Communications Act) which is not a decision specified in Schedule 8; (ii) a restriction or condition made in relation to the Electronic Communications Code; (iii) an emergency direction (made under section 132 of the Communications Act); and (iv) any direction with respect to the radio spectrum (under section 156 of the Communications Act) that is not about the making of a decision specified in Schedule 8.
100 [2002] CAT 8.
101 See, for example, *BetterCare Group v Director General of Fair Trading* [2002] CAT 6 and *Claymore Dairies Ltd* and *Express Dairies v Director General of Fair Trading* [2003] CAT 3.

caution has been taken by the regulators in framing any rejection of a complaint, particularly in circumstances where some preliminary form of investigation has been undertaken. Generally, a complainant to OFCOM whose complaint is not to be taken past the preliminary investigation stage will receive a short letter setting out OFCOM's intention not to undertake any further investigation on the grounds of administrative priority. This is based on European case law which has clearly stated that when regulators prioritise those cases they wish to deal with, this is an exercise of their administrative discretion and not a non-infringement decision[102]. Great care will be taken by OFCOM in framing this letter to ensure there is no sense in which it can be construed as reaching any findings on the merits or otherwise of the original grounds for complaint.

The CAT's judgment in cases such as *Freeserve* has also undoubtedly impacted on the process which regulators such as OFCOM have adopted in screening complaints (see Section 7.1.5 above).

9.3 Alternative Options Available to Parties who Wish to Litigate a Non-appealable Decision

In so far as a decision is not an appealable one, a party can only take action against OFCOM by seeking a judicial review of the exercise of its administrative discretion in the Administrative Court[103]. Otherwise, it will be necessary to determine whether there are other grounds by which they can enforce their legal rights, for example by launching ('direct') High Court litigation against the other party(ies) concerned. In the field of competition law enforcement (see 'issues box' at Section 7.1.5), this is increasingly an option which companies are being encouraged to consider as the regulators continue to step back from investigating complaints, albeit there are likely to be significant costs involved in persuading a court to reach its own finding of a competition law infringement.

9.4 Who is Entitled to Appeal?

It is self-evident that a party which is the subject of a decision should be entitled to exercise any rights of appeal against that decision in the CAT. However, there may be third parties with standing to appeal if, in accordance with the wording employed by section 192(2) of the Communications Act, they are a 'person affected' by a decision. The Competition Act 1998 also provides for third party rights of appeal if they can show they have 'sufficient interest' in the decision with respect to which the appeal is made[104].

More commonly, however, third parties will apply to the CAT under rule 16 of the Tribunal Rules 2003 to be able to intervene in an existing appeal—if granted, this application (which must be made within three weeks of publication of the notice of appeal on the CAT's website[105]) will entitle a party to have their views heard

---

102 This principle was confirmed by the European Court in Case T-24/90 *Automec Srl v Commission*.
103 For example in the case of *Cityhook v OFT* (Case No 1071/2/1/06), Cityhook sought judicial review of the OFT's decision to close its file on their complaint as well as appealing to the CAT. The CAT subsequently found that there had been no appealable decision taken by the OFT (see [2007] CAT 18).
104 Competition Act 1998, Section 47, as substituted by section 17 of the Enterprise Act 2002.
105 Rule 15(2)(f) of the CAT's Procedural Rules. The CAT enforces this deadline very strictly—see Truphone's failed application in *T-Mobile and others v OFCOM* [2007] CAT 31.

through participation in the written and oral stages of an appeal. The relevant test applied by the CAT to determine an intervention application is whether the party concerned has a 'sufficient interest in the outcome of the proceedings'[106].

9.5 Standard of Review Exercised by the CAT and Powers in an Appeal

Other than in relation to appeals of decisions pursuant to OFCOM's concurrent 'market investigation powers' under the Enterprise Act 2002 (which are undertaken on judicial review grounds), appeals against decisions made by OFCOM or the Secretary of State will be 'on the merits'. This standard of review is much more comprehensive than that adopted under judicial review proceedings and the CAT has full jurisdiction to find facts, make its own appraisal of economic issues and apply the law to those facts. Accordingly, the CAT's powers in an appeal are extensive. As set out under paragraph 3 of Schedule 8 to the Competition Act 1998, it may:

- adopt interim measures to protect a party to an appeal[107];

- confirm or set aside a decision;

- remit the matter to the decision maker;

- impose or revoke or vary the amount of a penalty[108];

- give such directions as the regulator would have taken (eg to end an infringement); or

- make any decision which the regulator could have made.

In practice, the way in which the CAT exercises its powers will depend on the circumstances of a particular case. Often where it quashes a decision made by a regulator, the CAT will remit the matter back to them for further consideration. For example, in the *Freeserve* case (see 9.2 above), the CAT remitted the issue as to whether BT was guilty of abusive pricing practices to Oftel (as then was) for it to re-investigate the matter[109].

Where it refers a matter back to OFCOM, the CAT may also include directions as to the manner in which the regulator's powers should be exercised. This is particularly relevant in the context of OFCOM's dispute resolution powers. *In T-Mobile, BT, H3, Cable & Wireless & Others v OFCOM,* the CAT remitted a number of disputes relating to wholesale mobile call termination rates back to OFCOM with specific directions as to how those disputes should be determined[110].

An interesting issue dealt with in the context of the *Floe Telecom* case was whether, on remitting a matter back to OFCOM, the CAT could also impose a timetable

---

106 See, for example, Vodafone's successful application to intervene in the *Floe Telecom* case at CAT [2004] 2.

107 The test applied by the CAT for the award of interim measures mirrors that adopted by the European Courts, see, for example, *VIP Communications v OFCOM* [2007] CAT 12 where VIP's application for interim relief in the form of a payment by T-Mobile of approximately £1.7 million and order that T-Mobile supply up to 4,000 SIMs to VIP was rejected.

108 Note that a penalty imposed under the Competition Act is automatically suspended pending the outcome of an appeal, although the CAT can order that interest is payable at the conclusion of a case.

109 *Freeserve.com plc v Director General of Telecommunications* [2003] CAT 5, [2003] CAT 202.

110 Order of the CAT dated 17 November 2008 in Case Nos. 1089/3/3/07, 1090/3/3/07, 1091/3/3/07 and 1092/3/3/07.

for its re-investigation of that matter. Having annulled OFCOM's original decision, the CAT was concerned at the time that proceedings had taken and imposed a deadline for OFCOM to reach a new decision in the case. OFCOM appealed to the Court of Appeal on the question of whether the CAT had been entitled to act in this way. The Court of Appeal held that the CAT's jurisdiction in an appeal was properly limited to deciding appeals before it and it did not have any more general statutory function of supervising regulators, for example, by imposing such time limits[111].

The CAT has also had occasion to use its powers under Schedule 8 to substitute its own finding for that of a regulator, albeit not to date in respect of OFCOM. In *VIP Communications Ltd v OFCOM*, the CAT confirmed that, if necessary, it would have had the power to substitute OFCOM's findings of a non-infringement by T-Mobile of the Chapter II prohibition under the Competition Act for its own finding of an infringement. In so doing, it rejected submissions made by T-Mobile that it had no jurisdiction to do so. The CAT noted that to accept T-Mobile's arguments would give no effect to the CAT's 'appeal on the merits' jurisdiction[112]. However, the CAT does not have 'carte blanche' to re-make a decision. This was evidenced in a (second) decision by the Court of Appeal in the *Floe Telecom* case where it held that although the CAT's overall decision (to uphold OFCOM's non-infringement finding) was correct, it had been wrong to also amend part of OFCOM's decision on the basis of deficient reasoning. In particular, the CAT had mis-used its powers to deal in its judgment with (additional) issues which had not been necessary to the disposal of the appeal against OFCOM's original decision[113].

9.6 Price Control References to the Competition Commission

As alluded to above, appeals raising 'price control matters' are treated differently under the Communications Act, the general principle being that such matters must (under Section 193) be referred to the Competition Commission. Any remaining aspects of an appeal will stay within the jurisdiction of the CAT. Price control matters are those set out under Section 193(10) and also in Rule 3 of the Competition Appeal Tribunal (Amendment and Communications Act Appeals) Rule 2004 (Tribunal Rules 2004), and relate to conditions imposing price control on operators with SMP:

- in relation to the provision of access to networks and relevant equipment (under section 87(9) of the Communications Act);

- in relation to the regulation of services for end-users of electronic communications networks (under section 91 of the Communications Act); and

- in relation to the hiring of telephones which are hardwired to an electronic communications network (under section 93(3) of the Communications Act)[114].

111 *OFCOM and OFT v Floe Telecom Ltd* [2006] EWCA Civ 768. See in particular the opinion of Lloyd LJ at para 34 of the judgment.
112 *VIP Communications Ltd v OFCOM* [2007] CAT 3, at paragraphs 45–46.
113 In the Floe Telecom case, the CAT has made findings as to the correct interpretation (under EC law) of Vodafone's Public Mobile Operator Licence which, although not determinative in terms of its overall judgment, were otherwise likely to impact on OFCOM's enforcement activities in the future—hence the reason for the appeal before the Court of Appeal. *OFCOM v Floe Telecom Ltd* [2009] ENCA Civ 47.
114 Rule 3(a)–(c) of the Tribunal Rules 2004 also specifies that, in order to be a price control matter for the purposes of Section 193, a dispute must in addition relate to: (a) the principles applied in setting out the condition which imposed the price control; or (b) the methods applied, calculations or data used in determining the price control; or (c) what the provision imposing the price control should be.

To date, each of the appeals dealt with by the CAT under these powers has concerned OFCOM's decision of 27 March 2007 to impose price controls on the UK MNOs in respect of wholesale voice call termination. One of those appeals was by H3G who argued, amongst other things, that the price control imposed on it was not an appropriate (or proportionate) response to OFCOM's findings of SMP. After a preliminary hearing, the CAT ruled that this particular aspect of H3G's appeal was not a 'price control matter' to be referred to the Competition Commission under section 193 – the question as to whether to impose a price control was separate from that as to what the price control should be[115]. H3G's arguments were subsequently dealt with in a separate judgment by the CAT concerning the non-price control matters raised by its appeal[116].

Where it makes a price control reference, the CAT will (after consulting with the parties concerned) set out those questions to be determined together with such directions as to the procedure that the Competition Commission should apply in reaching its determination. Ordinarily, the Competition Commission is to determine a price control reference within four months[117].

Once it makes a determination on a price control reference, the Competition Commission will notify the CAT which is then required to decide the matter on the merits in accordance with the Competition Commission's direction (Communications Act, section 193(6)). However, if the CAT considers the determination would be set aside on an application for judicial review, it can depart from the Competition Commission's findings (section 193(7)).

9.7 Appeals from the CAT

An appeal lies from a CAT decision on a point of law to the Court of Appeal or in Scotland to the Court of Session. Consent for an appeal to proceed must be obtained in the first instance from the CAT or, if it refuses to grant permission, directly from the court to which an appeal is made. There have been several occasions on which the CAT has refused permission to appeal but an appeal has proceeded on the basis of permission granted by the appeal court[118].

9.8 Costs

Rule 55 of the Tribunal Rules 2003 gives the CAT the power, at the conclusion of a case, to make such order as it sees fit in relation to costs. Unlike in civil litigation before the courts, the CAT does not apply the 'costs follow the event' principle and instead has generally been slow to make costs awards against individual parties unless their conduct has been manifestly unreasonable[119]. The CAT has also generally appeared willing to take account of the impact of any cost award against OFCOM on its ability to defend actions brought against it by large, well resourced companies[120].

---

115 *Hutchinson 3G UK Limited v OFCOM* [2007] CAT 27.
116 *Hutchinson 3G UK Limited v OFCOM* (*Mobile Call Termination*) [2008] CAT 11.
117 Tribunal Rules 2004, Rule 5. In the aforementioned *Mobile Call Termination* case the CAT extended this deadline to reflect the complexity of the issues at hand. The Competition Commission reached its final determination on 16 January 2009 (it had received the questions on 18 March 2008).
118 In *Floe Telecom* (footnote 76 above), the Court of Appeal granted permission to appeal following the CAT's decision to reject an application from OFCOM for an appeal.
119 See *Hutchinson 3G (UK) v OFCOM* [2007] CAT 7.
120 For an example of a cost award made against OFCOM, see *The Number (UK) Limited v OFCOM*, order of 28 April 2006.

Annex

Decisions Issued by OFCOM Under the Competition Act 1998

OFCOM maintains a list of all of its 'closed cases' under the Competition Act 1998, that is those cases where it has decided to open an investigation into allegations of anti-competitive behaviour and proceeded to a formal 'decision' (ie rather than rejecting a complaint for reasons of administrative priority—see Section 9.2 above)[121]:

Case Name (reference)	Date of decision	Synopsis	Outcome[122]	Appeal details
Disconnection of services by Vodafone in respect of Floe Telecom Limited (CW/0066/07/03) **NB: First decision in this case and was adopted by Oftel. Fresh decision adopted (by OFCOM) on 30 June 2005 (see below).**	3 November 2003	Floe Telecom Limited (Floe Telecom) was a provider of GSM gateway services. Those services use the Subscriber Identity Modules (SIMs) of a mobile network operator (MNO) on whose network a call is to be delivered to make a call from a fixed network appear to be a call from a mobile on the same mobile network so that it is charged at the MNO's lower retail rate for such 'on-net' calls. Floe Telecom claimed Vodafone had unlawfully refused to supply it with GSM gateway services and the SIMs necessary to provide those services	Non-infringement of the Chapter II prohibition Provision of the relevant services to Floe Telecom was unlawful as Floe Telecom did not have the necessary Wireless Telegraphy Act licences and therefore any refusal on the part of Vodafone was objectively justified.	Floe Telecom appealed to the CAT who set aside the disputed decision in a judgment dated 19 November 2004.[123] Subsequently, the CAT issued a direction that OFCOM must either issue a statement of objections against Vodafone (setting out its intention to find an abuse) or reach a new non-infringement decision within six months. OFCOM appealed this direction to the Court of Appeal who determined that the CAT had acted outside of its jurisdiction in imposing such a timetable on OFCOM.[124]

---

121 Available at http://www.ofcom.org.uk/bulletins/comp_bull_index/comp_bull_ccases/closed_c_act/. Note that a similar list of all closed investigations (ie including those under OFCOM's other powers of enforcement) is also available at OFCOM's website.

122 As set out above (see Section 7.A), in order to implement the 'Modernisation Regulation' (Regulation 1/2003/EC) into UK law, as of 1 May 2004 OFCOM was given the power under the Competition Act 1998 to apply Article 81 and 82 of the EC Treaty in full, in addition to the Chapter I and II prohibitions.

123 Case 1024/2/3/04 *Floe Telecom Limited (in administration) v Oftel* [2004] CAT 18.

124 *OFCOM and OFT v Floe Telecom Limited* [2006] EWCA Civ 768.

Case Name (reference)	Date of decision	Synopsis	Outcome[122]	Appeal details
Disconnection of services by T-Mobile in respect of VIP Communications Limited (CW/00662/07/03) **N:B: First decision in this case and was adopted by Oftel. Fresh descision adopted (by OFCOM) on 30 June 2005 (see below).**	31 December 2003	VIP Communications Limited (VIP) brought a complaint against T-Mobile on essentially the same grounds as that brought by Floe Telecom (see above), except that the relevant provider of GSM gateway services in this instance was T-Mobile not Vodafone.	Non-infringement of the Chapter II prohibition on the basis that refusal was objectively justified (for the same reasons as set out above in relation to Floe Telecom).	VIP appealed to the CAT who set aside the disputed decision and remitted it to OFCOM to re-consider the matter.[125]
Director Enquiries Advertisement on the cover of the BT Phonebook (CW/00604/03/03)	31 December 2003	Cable & Wireless (supported by a number of other companies) complained that BT was abusing its dominant position by using the cover of the BT Phonebook to advertise its new Director Enquiries number. The complainants alleged that BT was attempting to leverage its dominance in the paper directories market into the new market for 118 Director Enquiries services.	No infringement of Chapter II prohibition.	
Suspected margin squeeze by Vodafone, O2, Orange and T-Mobile (CW/00615/05/03)	26 May 2004	OFCOM own-initiative investigation into allegations that the mobile operators concerned were pricing the delivery of certain fixed-mobile calls for business customers at levels that constituted a margin squeeze vis-à-vis the wholesale charges that fixed operators paid for mobile call termination.	No infringement of Chapter II prohibition or Article 82 EC Treaty.	

125 Case 1027/2/3/04 *VIP Communications Limited v OFCOM*, Order of 1 December 2004.

Case Name (reference)	Date of decision	Synopsis	Outcome[122]	Appeal details
BT's residential call tariffs (CW/00760/03/04)	11 July 2004	OFCOM own-initiative investigation into BT's announced plans to withdraw its standard residential line rental and call charges and migrate existing customers paying the standard rates to 'BT Together Option 1', as well as a new pricing policy for the totality of its BT Together plans. OFCOM investigated concerns that carrier pre-selection operators (who provided competing retail services whilst relying on access to BT's physical lines) would, following these retail price changes, be unable to compete effectively for the provision of calls given the wholesale price they had to pay to BT for call conveyance.	No infringement of the Chapter II prohibition or Article 82 EC Treaty.	
BT 0845 and 0870 retail price change (CW/00647/07/03)	19 August 2004	OFCOM own-initiative investigation into BT's reduced retail charges for 0845 and 0870 numbers. Specifically, it investigated whether by reducing the out payments made to ISPs in order to pass on the effect of the lower retail charges, it was effecting a margin squeeze on its (ISP) competitors.	No infringement of the Chapter II prohibition.	

Case Name (reference)	Date of decision	Synopsis	Outcome[122]	Appeal details
Pricing of BT Analyst (CW/00718/11/03)	27 October 2004	Complaint by Designbyte Limited that BT, by offering BT Analyst (a desktop bill analysis software package) free of charge to corporate customers who used its Onebill service (which consolidated a business customer's separate bills for individual services into a composite invoice), was unlawfully bundling the two products together with the effect of excluding competing providers of the software.	No infringement of the Chapter II prohibition.	
BT Wholesale Limited: reduced rates for wholesale calls (CW/00802/11/04)	29 June 2005	Complaint by Gamma Telecom Limited (Gamma) that BT's announced reductions in wholesale calls tariffs represented an anti-competitive margin squeeze and, as such, an abuse of its dominant position. Gamma also sought interim measures in the form of an immediate reversal of BT's price reduction.	OFCOM rejected the application for interim measures on 22 April 2005. In its final decision, OFCOM found that no margin squeeze had taken place and therefore no abuse of BT's dominant position under the Chapter II prohibition.	

Case Name (reference)	Date of decision	Synopsis	Outcome[122]	Appeal details
Re-investigation of Floe Telecom (CW/00805/12/04) **NB: second decision following CAT judgment to remit original (Oftel) decision back to OFCOM**	29 June 2005	See original decision of 3 November 2003	Confirmed decision that Vodafone had not abused its dominant position under the Chapter II prohibition by disconnecting the SIMs used in Floe's VIP GSM gateways and refusing to supply Floe Telecom with any more SIMs.	Floe Telecom appealed to the CAT which found that, although OFCOM's decision contained inadequate reasoning in a number of areas (in particular in relation to its analysis of Vodafone's Public Mobile Operator Licence), those errors did not vitiate the conclusion that Vodafone had not abused a dominant position.[126] OFCOM and T-Mobile subsequently sought permission to appeal the part of the CAT's decision which considered the errors in OFCOM's reasoning to the Court of Appeal. That permission was refused by the CAT[127], but subsequently granted by the Court of Appeal.[128] The Court of Appeal handed down its judgment on 10 February 2009 in which upheld the parties' appeal and therefore reversed the CAT's findings as to the purported deficiencies in OFCOM's reasoning.[129]

126 *Floe Telecom Limited (in liquidation) v OFCOM* [2006] CAT 17.
127 *Floe Telecom Limited (in liquidation) v OFCOM* [2007] CAT 15.
128 Cases C3/2007/0658 *OFCOM v Floe Telecom Ltd* and C3/2007/0665 *T-Mobile v Floe Telecom Ltd.*
129 *OFCOM & Anor v Floe Telecom Ltd* [2009] EWCA Civ 47.

Case Name (reference)	Date of decision	Synopsis	Outcome[122]	Appeal details
Re-investigation of VIP Communications (CW/00806/12/04).	30 June 2005	See original decision of 31 December 2003 above.	Confirmed decision that T-Mobile had not abused its dominant position under the Chapter II prohibition by disconnecting the SIMs used in VIP's GSM gateways and refusing to supply VIP with any more SIMs.	VIP appealed to the CAT and also sought Interim Relief (reconnection to T-Mobile's network and the resumption of the supply of T-Mobile SIM cards) pending determination of its substantive appeal against OFCOM's original decision.\n\nThe application for interim relief was rejected in a judgment dated 27 February 2007.[130]\n\nVIP's substantive appeal was stayed pending final resolution of the *Floe Telecom* case (see above).[131] Final judgment in this case is still awaited.
BT's pricing of digital and cordless phones (CW/00828/03/05)	6 August 2006	Complaint by rival providers (claimed anonymity) that BT was abusing its dominant position in the price it set for various models of its branded cordless telephones. In particular, it was alleged that BT was selling at below cost (ie predatory pricing).	BT was not dominant in the relevant market (for fixed line telephones in the UK) and, in any event, its pricing behaviour did not constitute predatory pricing under the Chapter II prohibition.	

130 Case 1074/2/3/06 (IR) *VIP Communications (in administration) v OFCOM (Interim Measures)* [2007] CAT 12.
131 Case 1027/2/3/04 *VIP Communications (in administration) v OFCOM.*

Case Name (reference)	Date of decision	Synopsis	Outcome[122]	Appeal details
BBC Broadcast: provision of media access services (CW/00842/06/05)	4 June 2007	Complaint by Independent Media Support Group Plc (IMS) that BBC Broadcast (BBCB)'s agreement with Channel 4 for the provision of media access (eg subtitling, audio-description and signing) which enable hard of hearing/sight to watch or listen to TV programmes was (a) an anti-competitive agreement, and (b) constituted an abuse of BBCB's dominant market position. In particular, IMS relied on the length and exclusive nature of the contract concerned.	BBCB was not dominant in the relevant market and there was also no grounds for any Chapter I Competition Act 1998/Article 81 EC of the Treaty infringement on the basis that the relevant agreement did not appreciably restrict competition.	IMS appealed to the CAT which rejected the appeal in its entirety (judgment of 20 May 2008).[132] Permission to appeal to the Court of Appeal refused by the CAT (order of 24 July 2008) and further application to the Court of Appeal dismissed by Longmore LJ (judgment of 25 November 2008).[133]
BT's charges for NTS call termination (CW/00823/03/05)	1 August 2008	Complaint by Cable and Wireless (C&W) that BT's charges for termination of calls on the 0870, 0845 and 0820 ranges for the period from 1 May 2004 constituted an abuse of a dominant position. In particular, C&W argued the charges imposed an anti-competitive margin squeeze, discriminated in favour of BT's own business and were excessive.	No infringement of Chapter II prohibition in the Competition Act 1998 or Article 82 of the EC Treaty.	

132 Case 1087/2/3/07 *Independent Media Support Ltd v OFCOM* [2008] CAT 13.
133 *Independent Media Support Ltd v OFCOM & Ors* [2008] EWCA Civ 1402.

Index

[all references are to page number]

231